The *Coming* Revolution in Agriculture

by Dr. Harold Willis

Designed and produced by A-R Editions, Inc.
315 W. Gorham, Madison, Wisconsin

ISBN 0-912311-03-7

Printed in the United States of America

For sale by the author, P.O. Box 692, Wisconsin Dells, Wisconsin 53965

Dedicated to all
who seek answers to
difficult questions and
hope for the future.

Drawings and photographs
are by the author

CONTENTS

The *Coming* Revolution in Agriculture

by Dr. Harold Willis

Foreword

". . . the U. S. agricultural system has produced an increasing abundance of food and fiber, using relatively fewer resources. The unparalleled productivity gains in American agriculture have resulted in a broader choice of nutritious foods available for a declining portion of income." (from article, "Agriculture's Vital Role for Us All," by J. A. Zellmer & R. McFall Lamm, *Food—From Farm to Table*, 1982 Yearbook of Agriculture, USDA, p. 2).

These words of praise for modern American agriculture sound comforting, don't they? Recent statistics from the same Yearbook of Agriculture seem to reinforce the consumer's opinion that agriculture, for the most part, is doing well.

- Ten years ago, one farm worker supplied enough food and fiber for 47 people; he now supplies enough for 78, up from a mere seven persons at the beginning of the century (p. 2).
- Total agricultural output has more than doubled in the last 50 years, though the agricultural land base has not substantially changed (p. 2).
- Production of grains and oilseeds has expanded 2 1/2 times over levels of 30 years ago, and exports of these commodities have increased 10 times in the same period (p. 5).
- The United States now accounts for 55 percent of total world grain exports, over half of world soybean exports, and almost a third of cotton exports (p. 5).
- Rising agricultural exports have been a major factor in holding down the U. S. trade deficit in recent years. In 1981, agricultural trade produced a surplus of $26.5 billion, while nonagricultural trade experienced a deficit of $56.6 billion (p. 5).
- Today Americans spend less than 17 percent of family income on food (p. 2). In 1930 a worker had to labor 50.6 minutes to earn the income needed to buy a dozen eggs. In 1950, the figure was 28.3 minutes, while in 1980 it was only 8.9 minutes (p. 53).

The statistics can even be used to show the farmer how well off he is: "Individual farmers have gained in real terms, since each farmer and farmworker is receiving more consumer time for their efforts than before . . . Consumer time per unit of food output in recent years has been shared among a much smaller number of farmers than in 1930, so that the consumer time received per unit by each farmer and farmworker in

1970 and 1980 is more than 10 percent greater than earlier years. This represents a real gain in the well-being of the farmer" (from article, "Productivity and the Real Cost of Our Food," by L. D. Teigen, 1982 Yearbook of Agriculture, p. 53).

Aren't statistics wonderful? If you have one foot in a bucket of boiling water and the other in a bucket of ice water, on the average you're comfortable!

But ask farmers how well off they are and you'll get a far different picture! After several years of sharply declining net income, farm problems have finally become so bad that even the prime time network news programs are devoting time to agricultural subjects, and in all fairness, the recent USDA Yearbooks of Agriculture do not ignore the problems either.

The basic theme of the 1982 Yearbook of Agriculture (titled *Food—From Farm to Table*) is food and agriculture in these days of changing economics. A few quotes:

> "Farms were largely self-sufficient during much of American history. It wasn't until the early 20th century that farmers purchased more inputs from the nonfarm sector than were produced on the farm or by other farmers. Today nearly two-thirds of farm inputs, by value, are purchased from nonfarm suppliers" (p. 7).
>
> "The family farm remains foremost in images about farms and farming, and it continues to rank high in announced goals in the future. Yet it is gradually disappearing" (p. 12).
>
> "Demand conditions shifted sharply in 1981 and 1982 as economic activity slowed in the United States and around the world. These conditions combined with high crop production levels to bring about lower farm crop prices, lower farm income, and in turn lower farmland prices" (p. 19).

Did you notice that? High crop production levels helped to lower prices. Yet the basic theme of the 1981 Yearbook of Agriculture (titled *Will There be Enough Food?*) is just the opposite: food scarcity in the world as a whole. Notice these quotes:

> "The 1980's will be a decade of critical adjustments in world and U. S. agriculture. Foreign demand for food and agricultural products is likely to grow at a record or near-record pace . . . The world's dependence on the United States for agricultural supplies is likely to increase significantly" (p. 30).
>
> "In any case, American agriculture cannot feed the world (p. 8).
>
> "Over the next several years, prospects for food supply and equitable distribution of food are fair at best" (p. 48).
>
> "No longer is U. S. agriculture faced with chronic surpluses" (p. 42).

What was that!? No chronic surpluses? But I thought that was why 1983 brought the PIK program and a dairy surplus reduction program. Doesn't the right hand know what the left hand is doing?

In spite of all the country's economic woes, and those of agriculture in particular, we often hear of a rosy picture of the future—of exciting research now going on which it is hoped will open the door to unlimited food supplies, eliminate world hunger, and bring a new era of prosperity. Some of these areas of agricultural research include (1) genetic engineering to produce man-made varieties of plants and animals which will resist disease and drought, and produce greater yields; (2) improving photosynthesis efficiency to increase crop yields; and (3) nitrogen-fixation to allow all crop plants to obtain "free" nitrogen nutrients from the air.

All of these fields of research involve "tinkering" with living organisms and their processes. Thus, this hopeful research has been called "*the biological revolution in agriculture*." Even the big "chemical" companies are jumping on the bandwagon. Such corporate giants as DuPont, Upjohn, Shell, Allied Chemical, and Dow are buying up hybrid seed and nitrogen-fixing bacteria companies, giving large research grants, and doing their own research in these areas.

Is this the right direction to go? There IS a biological revolution coming in agriculture, but as time will tell, it may not be exactly what the researchers and visionaries have in mind.

One thing is certain, without a rejuvenation or revolution in agriculture, *modern agriculture is doomed!* Let's explore the exciting prospects for the future in agriculture.

This book follows on the heels of its predecessor, *The Rest of the Story . . . about Agriculture Today*. That book pointed out many of the problems of modern agriculture, their causes and solutions, in a general way. A reading of the previous book would provide an excellent background for this one. What we intend to do here is cover some of the topics of *The Rest of the Story* in more detail and to go a step beyond into relatively new and uncharted territory—high tech and the future. It should be a fascinating journey!

CHAPTER 1
The Fruits of Modern Agriculture

"You shall know them by their fruits." Matthew 7:16

What is "modern agriculture"? Is it the gigantic tractors and plows pictured in the glossy pages of the farm magazines? The latest bug- and weed-zapping chemicals proclaimed on prime-time TV? Is it no-till farming, in the name of soil conservation? Is it 200 bushel per acre corn and supercows that produce 20,000 pounds of milk per year? Yes, it is all that and more. To read the advertisements, plus the optimistic predictions of what science and technology will do for us tomorrow, you would think that we have nothing to worry about. Is everything coming up roses? Or don't roses also have a lot of thorns?

Let's look at modern agriculture, how it developed, how it operates, and where it is headed.

What is it? What we call modern agriculture is the methods and systems that are an outgrowth of various scientific and technological developments which have become widespread in the last forty years; that is, since World War II. Among the manifestations of modern agriculture are (1) large-scale use of inorganic salt fertilizers, herbicides and other pesticides, and growth stimulants and drugs (for animal agriculture); (2) increased size of farms and fields, with large machinery, the clearing of windbreaks, irrigation, and concentrated animal feeding operations; and (3) improvement of plant and animal characteristics by hybrids, strain improvement, and artificial breeding.

These technological advances began to become widespread from 1941–1945, and have "exploded" since then into what historians call the Second American Agricultural Revolution.*[1] They have truly spawned a revolution—they have transformed agriculture into a "whole new ball game."

*The First American Agricultural Revolution began during the Civil War (1861–1865) when increased demand for food led to the commercialization of agriculture in the northern states.

[1]W. D. Rasmussen, *Readings in the History of American Agriculture*, 1960, p. 275, 309.

However, these technological developments did not spring into being full blown during World War II. They had their predecessors; they were developed gradually over the decades before. Let's look at how some of them originated.

Fertilizers. Knowledge that adding certain natural materials to the soil (such as lime, gypsum, and manure) improves yields has been around since ancient Roman times and before. The first European settlers in Jamestown in 1609 and the pilgrims in 1621 learned from the Indians how to grow corn (maize) by adding a dead fish to each hill. The impetus that gave rise to the modern synthetic fertilizer industry came from the research of the German chemist, Justus von Liebig. After studying plant growth in relation to the amount of materials used up from the soil, he concluded that certain amounts of minerals were needed to produce a certain amount of growth or yield and that a total diet of inorganic minerals could supply the plant's needs. This gave rise to the current "balance-sheet" theory of plant nutrition in which fertilizer amounts are recommended according to the farmer's yield goal, with little or no attention being paid to the contribution of organic matter, nitrogen-fixation, and other beneficial effects of soil organisms (see Chapter 2 in this book and Chapters 3 and 6 in *The Rest of the Story* for more details on the value of soil life).

In the U. S. A., mixed chemical fertilizers first became commercially available in 1849, but the use of lime and synthetic fertilizers simply did not become widespread until World War II, probably partly because farmers previously farmed in harmony with natural systems, but also because World War II brought an incentive to greatly increase

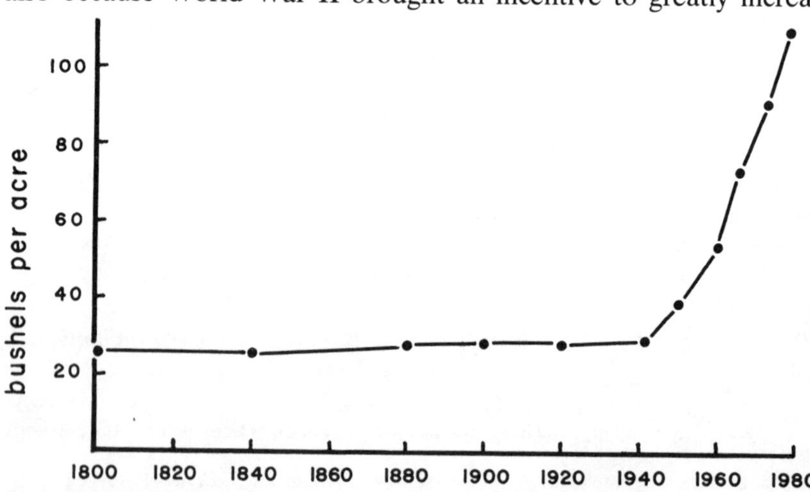

Increase in U. S. corn yields since 1800 (Source, USDA statistics).

yields. More food was needed to help make up for lost production in Europe, and the prices farmers received doubled. The annual use of "N-P-K" fertilizers increased 95% in 1945 over the 1935–39 period, while lime usage increased three times.[2]

Since then, the use of synthetic fertilizers and crop yields have climbed astronomically. Commercial fertilizer used in 1980 amounted to over 75 million tons, compared to slightly less than 21 million tons in 1950.[3]

Pesticides. Just as with fertilizers, the use of relatively safe insecticides and other pesticides has existed for centuries. About 1000 BC, the Greek author Homer mentioned the use of sulfur for pest control. By 900 AD the Chinese were using arsenic to control garden pests. By the late 1600's, pesticides were used in Europe and later in the U. S. A. Both inorganic chemicals (sulfur, arsenic, mercuric chloride, carbon disulfide, etc.) and natural botanical substances were used (tobacco, pyrethrum, quassia, derris, etc.). The first really important synthetic organic hydrocarbon, DDT, was first made in the laboratory in 1874 in Europe. It was found to be a good insecticide and was used during World War II for human louse and mosquito control. With favorable publicity, its use increased greatly (over four times) after the war, in agriculture.[4] Since then the genie of chemistry has truly been unleashed. Literally thousands of new chemicals are synthesized and tested every year.

Mechanization. Modern farming would not be possible without the many and varied machines used for plowing, planting, cultivating, and harvesting. Early milestones of mechanization include the invention of the cotton gin by Eli Whitney in 1793, the steel plow manufactured by John Lane in 1833, the horse-drawn grain reaper by Cyrus McCormick in 1834, the grain combine by J. Hascall in 1836, the steel share plow by John Deere in 1837, a practical threshing machine by the Pitts brothers in 1837, and the horse-drawn double-row corn planter by D. S. Rockwell in 1839. The first steam tractors were produced in the U. S. as early as 1849, and the first gasoline tractor was built in 1892 by John Froelich. The first gas tractor factory began in 1903, but tractors did not really completely displace animal power until around World War II (it wasn't until about 1950 that there were more tractors than horses on American farms).[5]

Hybrids. Farmers and herdsmen have long improved their crops

[2]W. D. Rasmussen, *ibid.*, p. 275, 279.

[3]*Agricultural Statistics*, USDA, 1981, p. 466; *Statistical Abstract of the U. S.*, 1960, p. 640.

[4]H. H. Shepard, *The Chemistry and Action of Insecticides*, 1951, p. 4, 303–304.

[5]W. D. Rasmussen, *ibid.*, p. 159–160, 278, 297–305.

and animals by selecting the strongest or best-producing individuals for producing future generations. In 1812, John Lorain in Pennsylvania demonstrated that crossing two varieties of corn resulted in greater yield. By 1850, many farmers all over the country were purposefully crossing different varieties. But the principles of genetics were not discovered scientifically until 1866, when the Austrian monk, Gregor Mendel, published his experiments with garden peas in an obscure journal. But Mendel's work went unnoticed by the scientific world until 1900. In the early 1900's the concepts of genes and mutations were announced by Thomas Hunt Morgan and Hugo DeVries, respectively. Also, in 1900, W. A. Orlon and E. L. Rivers developed a variety of cotton resistant to cotton wilt disease. From 1916 to 1918, Donald F. Jones developed modern corn hybrid methods (double crossing, using four inbred lines instead of two). In 1926, hybrid seed corn became commercially available, and it gradually displaced the use of open pollinated (non-hybrid) varieties, especially after World War II.[6]

So we can see from these examples that modern agriculture resulted from a natural growth process that began many years earlier, and that it has "exploded" since World War II along with the rest of science and technology. The sharply increasing world population curve greatly added to the demand for food and fiber, and science stepped in to fulfill that need.

Spreading the gospel. The present state and federal agricultural institutions can be traced back to as early as 1776, when proposals for an agricultural branch of the government were first made. President George Washington recommended the creation of a national board of agriculture in 1796. Nothing was done until 1839, when Congress appropriated $1000 of Patent Office fees for collecting agricultural statistics, conducting agricultural investigations, and distributing seeds, organized under an Agricultural Division of the Patent Office. In 1856, a 5-acre test garden was begun. On May 15, 1862, President Lincoln signed a bill creating an independent Department of Agriculture, headed by a Commissioner responsible to him. The Department was raised to cabinet status in 1889. The Morrill Land Grant Act of 1862 provided for the establishment of the state land grant colleges. The first state agricultural experiment station was established in 1875, the Connecticut Experiment Station (at Wesleyan University). The Hatch Act in 1887 provided a yearly federal grant to support an experiment station in each state.

In 1892, the cotton boll weevil invaded the U. S. from Mexico and in ten years the U. S. cotton industry was in danger of going under. Re-

[6]*Ibid.*, p. 60–61, 162, 215, 243.

search carried out by the U. S. D. A. developed control measures involving planting so that the crop matured early and escaped most weevil attack, but few farmers adopted them. S. A. Knapp of the Bureau of Plant Industry conceived the idea of getting some farmers to try new methods on their own farms (under an indemnity fund to reimburse a volunteer for any losses suffered). This succeeded so well that in 1906, the General Education Board of New York City agreed to supplement government funds to employ field agents to work full time with farmers in one county. Thus was born our system of county agents, supported by government and state university research.[7] In more recent years, considerable grant money has been given to universities to do research by the industries manufacturing agricultural products (fertilizers, pesticides, equipment, hybrid seeds, etc.) (see Chapter 7, *The Rest of the Story*). And of course, the whole agricultural journalism and advertising industry has grown up to disseminate information and promote products.

Marketing/distribution. At least as far back as the tin can (1818) and condensed milk (1856), there has been processed food, and the present marketing/processing/distribution system has developed since then. Other milestones along the way include the invention of the refrigerator in 1801 by Thomas Moore; efficient railroad transportation, which began to be developed in the 1840's; railroad refrigerator cars in 1851 and especially after 1868; the passage of pure-food laws by Illinois, Michigan, New Jersey, and New York in 1881; the first USDA export food inspection (pork products) in 1890; approval of the federal Pure Food and Drug Act and Meat Inspection Act in 1906; the Cotton Futures Act, approved, 1914; the first attempt to regulate marketing of farm products, the fixing of a minimum price for a bushel of wheat ($2.20) by President Wilson in 1917; the approval of the Grain Futures Act in 1922; approval of the Agricultural Adjustment Act in 1933; and approval of acts incorporating the principle of flexible price support, change in the parity formula, and authorizing set-asides and subsidies in 1948–49 and 1954.[8]

World-wide system. Such are the beginnings of our present system. No farmer in this country is untouched by it. Most countries in the world have dealings with it, whether by importing our agricultural products or exporting theirs or buying machinery or growing "miracle grains" of the Green Revolution.

Basic approach. But what are the fruits of modern agriculture? Are they all good—or are there undesirable side-effects? The basic philosophy or approach of this "scientific" system of agriculture seems to be to

[7]*Ibid.*, p. 105–106, 143, 178–179.
[8]*Ibid.*, p. 299–311.

manipulate and control nature, to change natural systems to suit (and profit) man.

Is this approach good? Now it is true that in the Garden of Eden, God instructed man to "subdue" the earth and "have dominion over" the living organisms in nature (Genesis 1:28), but it is also true that there are established laws by which the natural world operates, which cannot be broken with impunity.

• The over-all ecological food chain operates with the scores of species of plants, animals, and microbes being in a delicate balance, the so-called "balance of nature." A disturbance of one level or species in the food chain affects other links in the chain.

• The various organs of a person's, animal's, or plant's body have definite functions, and under proper conditions they operate well, and the organism is healthy. Improper nutrition and environmental conditions can cause malfunctions, allowing disease microbes, parasites, and other pests to invade.

• Any individual cell of a person's, animal's, or plant's body is made up of dozens of parts (nucleus, genes and chromosomes, ribosomes, mitochondria, endoplasmic reticulum, lysosomes, etc.), plus thousands of enzymes, energy-carrying, and "raw material" molecules. In a normal, healthy cell, all of this biochemical "machinery" works beautifully, performing the flurry of activities of any cell: breaking down food to produce energy, putting together components in growth and secretion activities, and in green plant cells, photosynthesis.

• The thousands of chemical molecules, ions, and atoms inside living organisms, in the non-living environment (rocks, soil, water, air), and in the whole universe, have definite properties (see any chemistry or physics textbook). This accounts for the distinct differences among, for instance, steel, paper, and water. And think of the different properties of ice, steam, and liquid water, all forms of H_2O.

Natural laws cannot be changed—or broken—without a penalty. The law of gravity, the speed of light, and the properties of oxygen simply can't be violated or changed. Science and technology in general, and the educational system they nurture, approach problems and develop new concepts with the philosophy of overpowering and changing nature to suit their purposes. If a mountain is in the way—move it! If an epidemic of grasshoppers is devouring crops—zap them with a poison! If corn won't grow on out-of-balance soil—develop a new variety that will!

Fruits. So with this in mind, let's look at some examples of the legacy of modern agriculture—the fruits of our "scientific" age of technology. Let's see if the results are desirable and for the good of mankind. To cite a couple of old sayings, "The proof is in the pudding" and "Nothing succeeds like success." How successful is modern science and its control-nature philosophy?

Vanishing soil. Soil is absolutely necessary for agriculture—fertile soil! Notwithstanding futurists' predictions (dreams) of hydroponic farming, the world is going to be fed with plants grown in soil for the foreseeable future. Some parts of the world were blessed with incredibly fertile soil, especially the United States of America. Crop yields in the days of the pioneers were every bit the equal of today's. According to the *Farmer's Every-day Book*, published in 1851, "Mr. J. P. Jones, of Sullivan Co., N. Y., in 1849, [raised] over 195 bushels of ears [of corn] per acre," while "Mr. Young, of Kentucky, in 1840, raised over 190 bushels per acre . . . in 1841, B. Bradley, Bloomfield, N. Y., raised 232 bushels on two acres" (p. 406–407).

But such yields did not always continue. Wrong farming practices often wore out land in a few years, so the family just packed up and moved west. Some farmers even spoke proudly of how many farms they had "farmed out" in their lifetime. But then by 1890, there was no more frontier—no more large areas of virgin land. Increasingly worsening soil loss in the late 1800's and early 1900's led to localized interest in soil conservation.[9] World War I took over 50 million acres of European land out of production, so Americans patriotically plowed 40 million acres of the Great Plains while market prices rose. Wheat became a cash crop. New machinery produced more efficiency and higher yields, but after the war this led to a glut on the market and lower prices. So the "natural" thing to do was plow more land to grow more wheat to pay off debts.[10] All this unprotected land combined with devastating droughts in 1933 and 1934 spawned the dust bowl, when the sun was darkened as far east as Boston.[11] This calamity plus the Great Depression produced a plethora of government New Deal agencies, including the Soil Erosion Service in 1935 and later that year the Soil Conservation Service.[12] So began the government's predilection for paying farmers to do what the government wants—to install terraces, to grow soil-conserving crops, to not grow "soil depleting" crops, or to not grow anything at all! Today there are grants available to persuade farmers to switch to "conservation tillage."

Surely all those conservation programs and all that money have reversed the loss of our precious soil. No doubt they did for a while, but what do we have today? Spurred on by some of the same economic fac-

[9]*Ibid.*, p. 265.
[10]W. Jackson, *New Roots for Agriculture*, 1980, p. 76.
[11]T. A. Bailey, *The American Pageant*, 4th ed., 1971, p. 881–882.
[12]W. D. Rasmussen, *ibid.*, p. 265.

tors as earlier in the century, farmers have plowed fencerow to fencerow—and then ripped out the fences—and windbreaks—to make room for gigantic tractors and center-pivot irrigation rigs as the topsy-turvy economy dictates "efficiency" and large acreage. Over-reliance on and overapplication of humus-destroying and soil life-destroying synthetic fertilizers and pesticides have made soil sterile and like powder or concrete.

The results are astounding. The Soil Conservation Service estimated (1979) an average yearly erosion loss of 9 tons per acre, but on rolling Iowa cropland, about 40–50 tons per acre, and on unprotected hillsides, 200 tons per acre are going down our rivers![13] Under ordinary agricultural conditions, new topsoil can form at a rate of 1.5 tons (a layer 0.01 inch thick) per acre per year.[14] Obviously this cannot go on much longer. In some areas, the topsoil is already gone and farmers are farming subsoil. The water-borne soil clogs up our lakes, reservoirs, harbors, and ship channels, requiring millions of dollars of dredging costs.[15] Dust bowl conditions have again started to develop in the Great Plains. In 1977, about 1 billion tons of soil were blown off of plowed fields.[16] In 1979, 1.4 million acres were damaged by wind erosion in 10 Great Plains states.[17]

Not only is our soil disappearing down our rivers and through the air, but nearly 7000 acres *per day* are being paved and gobbled up by creeping urbanization (about 2.5 million acres per year). An average traffic interchange takes 40 acres. Since 1945, an area larger than the state of Nebraska has been lost. About half of the time, cities and roads are built on good land, since there are fewer costs for leveling and grading.[18]

There have also been shocking losses in soil structure and fertility. The USDA estimates that in areas where the topsoil is 10–15 inches deep, the loss of 6 inches of topsoil may reduce crop yields from 20 to 42 percent.[19] According to a UPI news release of June 5, 1984, a United Nations report, "Harvest of Dust," released after a two-year study, warns of a worldwide threat from "desertification," the increase of deserts. The report states that a total of 8.6 billion acres of the world's crop and pasture land are being turned into arid deserts because of "bad farming and grazing practices, improper irrigation, and forest destruc-

[13]W. Jackson, *ibid.*, p. 19.
[14]D. Pimentel, *et al.*, *Science*, 1976, vol. 194, p. 150.
[15]L. E. Wright, *Yearbook of Agriculture*, USDA, 1981, p. 144.
[16]B. Gibbons, *National Geographic*, 1984, vol. 166, p. 355.
[17]J. Risser, *Smithsonian*, 1981, vol. 11, no. 12, p. 124.
[18]D. Pimentel, *et al.*, *ibid.*, p. 149.
[19]L. E. Wright, *ibid.*

tion." This is an area about the size of North and South America combined! Africa and Asia are the hardest hit, but the problem is increasing in other areas, including the United States.

Adequate humus (from 3–10%) is necessary to give soil a loose, crumbly texture (allowing entrance of water and air, reducing erosion, and making soil easy to work), to serve as a storehouse of vital plant nutrients, and to buffer soil against toxic substances and extreme pH changes. Humus is produced by a host of beneficial soil organisms (bacteria, actinomycetes, fungi, worms—see Chapter 2). Besides producing humus, some of them assist plant roots in absorbing nutrients (mycorrhizae), others provide additional nutrients (nitrogen-fixing bacteria, mineral-releasing bacteria), some supply plants with growth-stimulating substances (hormones and vitamins from bacteria and fungi), and some actually protect plants from diseases and pests (by producing antibiotics and by suppressing or destroying harmful "germs," worms, and insects).

Yet some commonly used fertilizers (anhydrous ammonia and high-salt fertilizers), as well as most pesticides and herbicides, kill, inhibit, or disturb the natural balance of this volunteer work force in the soil. High ammonia fertilizers actually make humus soluble so it can readily leach away (see Chapter 3). Humus cannot be produced if the "humus factory" is destroyed (soil life). Also, the tendency away from most farms having livestock toward all-grain farming (and over-concentration of

A herbicide spill.

livestock and poultry in large feeder operations) means there is little or no manure being recycled back into the soil to provide adequate humus (recycled crop residues do help, but animal manures used in addition are much better). Continual removal of crop matter (grain, stalks, hay) and insufficient replacement means that humus levels will soon drop too low to provide adequate soil fertility. Science and the Von Liebig mentality say that the lacking fertility can be made up totally by inorganic synthetic fertilizers. But many biochemical studies have shown that this simply is not true. Yields in bulk (quantity) may stay high, but *quality* and nutritional value of crops suffer immediately, and even bulk yields will eventually decrease, at least after it becomes economically prohibitive to keep pouring on the chemicals at increasing rates (see *The Rest of the Story*, Chapter 17).

Floods. Did you realize that topsoil and humus deterioration have been a major contributing factor in the devastating flooding that has occurred in recent years? In their book, *The Flood Control Controversy*, L. B. Leopold and T. Maddock, Jr. say (p. 59): "The incorporation of humic materials into the upper horizons [of the soil] and the consequent association of soil organisms promote the development of soil structure and good tilth which make the soil more permeable. Furthermore, organic material at the soil surface physically aids in diffusing water over the ground in numberless microchannels, and then, by preventing high velocity of flow, inhibits the cutting of rills or channels. This action on velocity increases the infiltration opportunity and also contributes to detention storage during runoff."[20]

Another quote: "Deterioration of tilth and structure lead to reduced water infiltration and water-holding capacity, reduced natural fertility, and increased erosion rates. If measures are not taken to restore the soil to its proper structure, tilth, and organic matter, their loss will have a snowball effect. Soil fertility and the infiltration rate will decrease, and organic matter content will decrease even more."[21]

According to USDA figures, soil with a 4 to 5% organic matter content can absorb a 4–6 inch rainfall in an hour, while soil with a 1 1/2–2% organic matter content can only absorb a 1/2–1 1/2 inch rainfall. Typical U. S. corn belt soils are down to only 0.5–2.5% organic matter.[22]

Just about any soils textbook will have a graph showing what happens to soil organic matter under typical agriculture. Organic matter is lost, at first rapidly, then more slowly. Tests at the Missouri Agricultural Experiment Station found organic matter losses of about 25% the first

[20]L. B. Leopold & T. Maddock, Jr., *The Flood Control Controversy*, 1954.
[21]L. E. Wright, *ibid.*, p. 146.
[22]C. Walters, Jr. & C. J. Fenzau, *An Acres U. S. A. Primer*, 1979, p. 95.

20 years of cultivation, 10% the next 20 years, and 7% the third 20 years.[23]

S. G. Archer says it eloquently in his book, *Soil Conservation* (p. 199–202):

> "The nation's river systems are rich, and some are dying. The trouble stems from man's use of the land. Virgin forests and verdant native grasslands that once protected mountains and plains are gone or are badly depleted. The protective forest canopy, with the leaf mold beneath, and the tall luxuriant grass protected the soil from the devastating power of the falling raindrops and held runoff water back until most of it soaked into the ground, protecting the soil while the surplus water moved slowly off to clear streams. That was the situation before the white man had dominion over the soils of this continent.
>
> "Today, fifty to three hundred years after man began to conquer the land with axes, plows, sheep, and cattle, there is little natural protection for the soil, and there is even less planned protection established by the most successful group of woodsmen, farmers, and ranchers the world has ever known. The Midas touch of the conquerors of the land turned natural resources of grass and trees into money.
>
> "But the land is bare.
>
> "Many of the nation's forests have been ravished, and most of the rest have been impoverished. The forest land produced crops for a while, and then gullies, where torrents of muddy water, unhindered, carried the remaining soil to creeks and rivers. The grasslands were either plowed or overgrazed, often with the same erosive effects. But even where erosion did not develop, the grass lost its ability to hold back the rainfall and to let it soak into the ground.
>
> "Cropland too often is tilled, planted, and harvested with never a thought of protection or of the damage being done to the structure of the soil. The rain packs the unprotected soil instead of enriching it, thus permitting less and less of the rain to go into the soil and more and more to run off, carrying with it huge quantities of topsoil to be deposited on the bottom lands and in the channels of creeks and rivers.
>
> "As man's dominion of the land became more complete, floods and flood damage increased in frequency and intensity.
>
> "Great avalanches of mud move into and down all of the nation's river systems, carried by more and more water as the soils, the ranges, and the forests deteriorate.
>
> "Rivers and creeks which once carried a steady flow of water throughout the year become raging torrents during heavy rains and then dwindle to trickles during the dry months, because the rainfall runs off instead of soaking in to renew the underground watercourses which used to feed the streams during the dry season.
>
> "The loss of lives in floods is recorded as news. Damage to property, especially in cities, makes the headlines when a great river goes on a rampage. Losses of crops are sometimes recounted, but the damage to the land by scour and silt deposition is seldom mentioned. In newspapers,

[23]H. D. Foth & L. M. Turk, *Fundamentals of Soil Science*, 5th ed., 1972, p. 135.

floods are generally attributed to big rains upstream, but the contributory factors of range, forest, and cropland deterioration are not recognized."

Yields. USDA studies have shown that loss of topsoil reduces crop yields. The loss of an inch of topsoil can reduce corn yields by 3 to 4 bushels per acre, and if the topsoil is only 10–15 inches deep, the loss of 6 inches can reduce yields by 20–42%.[24]

Beginning in the early 1970's, yields of most U. S. crops hit a peak, and then began falling, dipping up and down in succeeding years, or leveled off. Corn yields peaked in 1972 at 97.1 bushels/acre, then fell to 71.3 in 1974 (because of widespread drought) before inching back up to 91.5 in 1977. Wheat yields peaked at 33.9 bushels/acre in 1971 and have been pretty stationary at less than 31 since then. The experts do not know what is causing yields to stop growing at the rates they had been since the 1940's. Sylvan Wittwer of Michigan State University speculates that yields may be faltering because soils are losing organic matter, or are being compacted by heavy machinery, or because air pollution is harming crop growth, or because fewer pesticides and high-priced fertilizers are being used, or because less-productive land is being cultivated.[25]

Water too. An acre of corn requires about 500,000 gallons of water during a growing season. Agriculture accounts for 96 percent of the water consumed in the U. S. About 12% of U. S. cropland is irrigated. [26] In the water-short western states, irrigation uses about 80% of all water used.[27] We are "mining" water at such a great rate that the mammoth Ogallala aquifer, an underground sea of water stretching from South Dakota to the Texas panhandle, is dropping about 10 feet a year, while the groundwater is being replenished at only about 3 inches a year.[28]

Another problem associated with irrigation in dry climates is the build-up of salts at or near the surface because of high water evaporation rates from the wet soil. Also, the salinity of some rivers used for irrigation is increasing, such as the Colorado River, whose water is 18 times saltier near its mouth than at its headwaters.[29] Many acres of once-lush farmland have been turned into a salt desert.

Chemicals to the rescue! The growth of agricultural chemicals has been phenomenal. In 1940, only about 8.55 million tons of commercial fertilizer were used in the U. S., while in 1965, nearly 33 million tons were used.[30]

[24] L. E. Wright, *ibid.*, p. 144.
[25] *Farm Journal*, February 1978, p. 32, 34.
[26] W. Jackson, *ibid.*, p. 29.
[27] D. Pimentel, *et al.*, *ibid.*, p. 153.
[28] W. Whyte, *Yearbook of Agriculture*, USDA, 1981, p. 139.
[29] W. Jackson, *ibid.*, p. 30.
[30] *Statistical Abstract of the U. S.*, 1939, 1967.

Chemicals to control crop and hay insects costing $306 million were used on 42 million acres, and herbicides costing $356 million were used on 89.9 million acres in 1969, while in 1978, treated acreage increased to 75.7 million acres for crop and hay insects and 163.5 million acres for weeds. This doesn't include millions of dollars spent for chemicals to treat livestock and poultry, nematodes, crop diseases, and for defoliation ($121 million in 1969 and $166 million in 1974).[31] By 1978, about 1 billion pounds of pesticides were used annually in the U. S. About 51% were herbicides, 35% insecticides, and 14% fungicides. [32] In 1936, only about 30 pesticides were registered in the U. S., while by 1971, over 900 were registered, and pesticide usage is still expanding rapidly.[33]

Pesticide usage is varied. About 50 percent of insecticides is used on the nonfood crops of cotton and tobacco, while of the food crops, corn, fruits, and vegetables receive 33 percent. Forty-five percent of all herbicides is used for weeds in corn, by far the largest amount for any crop. Fruit crops consume the greatest amount of fungicides, 60 percent. On an acreage basis, omitting pastureland, 34 percent of U. S. cropland is treated with herbicides and 12 percent with insecticides. [34]

Results? What have been the results of this outpouring of mainly synthetic chemicals on the land? "Currently an estimated 33% of all crops is lost annually to pests (13% to insects, 12% to pathogens, and 8% to weeds) in spite of the use of pesticides and nonchemical controls."[35] Crop losses due to insect pests have increased nearly twice from the 1940's to 1974 (from 7% to about 13%), in spite of a ten-fold increase in insecticide usage. In the same period, losses from plant diseases and nematodes increased slightly (from 10.5% to about 12%), while losses from weeds declined (from 13.8% to 8%).[36]

"The substantial increase in crop losses due to insect damage despite increased insecticide use can be accounted for by some of the major changes that have taken place in agriculture since the 1940's. These include the planting of some crop varieties that are more susceptible to insect pests; destruction of natural enemies of certain pests, which created the need for additional pesticide treatments; increase in pesticide resistance in insects; reduced crop rotations and crop diversity—with an increase in continuous culture of a single crop; reduced FDA tolerance and increased 'cosmetic standards' of processors and retailers for fruits

[31]*Statistical Abstract of the U. S.*, 1980, 1981.
[32]D. Pimentel, *et al.*, *Bioscience*, 1978, vol. 28, p. 772, 778.
[33]F. L. McEwen, *Bioscience*, 1978, vol. 28, p. 775.
[34]D. Pimentel, *et al.*, *ibid.*, p. 778.
[35]*Ibid.*
[36]*Ibid.*, p. 778–779.

and vegetables; reduced sanitation, including less attention to the destruction of infected fruit and crop residues; reduced tillage, with more crop remains left on the land surface; culturing crops in climatic regions in which they are more susceptible to insect attack; and use of pesticides that have been found to alter the physiology of crop plants, making them more susceptible to insect attack."[37]

Read over that paragraph again. It's important. What Dr. Pimentel and his associates are saying is an indictment against many of the methods of modern agriculture, most of which are designed to produce greater yields more efficiently ("bigger is better") and to control nature.

What else? And what other side effects have there been as a result of increased use of agricultural chemicals? Few people are unaware of the growing menace of environmental pollution and the upset ecosystem. A few examples will suffice (see *The Rest of the Story*, Chapter 16 for more):

1. In 1974, the Environmental Protection Agency estimated that "as many as 14,000 individuals may be non-fatally poisoned by pesticides in a given year," with about 200 fatalities annually.[38]

2. Because of their widespread use, pesticides are consumed by most people. An EPA study found from 93 to 100% of the people tested had one or more pesticides in their bodies.[39] About 50% of foods sampled had detectable residues of insecticides.[40] "Unfortunately, little is known about the effects long-term, low-level dosages of pesticides may have on public health. Furthermore, the possible interaction between low-level dosages of pesticides and the numerous drugs and food additives the public consumes has not been fully studied."[41]

3. Chemical control for U. S. crop production was estimated in 1978 to be $2.2 billion annually.[42] But this does not include what Dr. Pimentel calls "external costs" of at least $3 billion annually. These include (a) hospitalization costs for 6000 human pesticide poisonings, (b) about 60,000 days of work lost from these pesticide poisonings (money costs of about 200 deaths were not included), (c) medical costs for an additional 8000 human pesticide poisonings treated as outpatients, (d) about

[37]*Ibid.*

[38]*Strategy of the Environmental Protection Agency for Controlling the Adverse Effects of Pesticides*, Office of Pesticide Programs, Office of Hazardous Materials, Washington, D. C., 1974.

[39]Human Monitoring Program, Environmental Protection Agency, *Pesticide Monitoring Quarterly Rept.* no. 6, 1976.

[40]Pesticide Residues in Food, p. 334–364 in *Environmental Pollution by Pesticides*, 1973, C. A. Edwards, ed.

[41]D. Pimentel, p. 62 in *Pest Control Strategies*, 1978, P. de Jong, ed.

[42]D. Pimentel, *et al.*, *Bioscience*, 1978, vol. 28, p. 772.

30,000 days of work lost from people not ill enough to be hospitalized, (e) about $12 million in direct honey bee losses, (f) reduced fruit crops from reduced pollination because of destruction of honey bees and wild bees, (g) livestock losses, (h) commercial and sports fish losses, (i) bird and mammal losses, (j) destruction of natural enemies of pests, resulting in outbreaks of other pests, (k) pest problems resulting from pesticide effects on the physiology of crop plants, and (l) increased pesticide resistance in pest populations.[43]

4. Over-use of commercial fertilizers and poor manure management, plus high levels of soil erosion mean that considerable amounts of fertilizer nutrients wind up in our rivers and lakes. These nutrients stimulate the growth of water plants (algae, "water weeds"), which then undergo a "population explosion," (called a bloom), sometimes clogging commercial and sporting waterways. And when they die, their decomposition uses up the water's oxygen and produces toxins which kill fish and other aquatic life. This process of increased growth and fertility of aquatic habitats is called eutrophication. It has become a serious problem since the late 1940's. For a while, it looked as if Lake Erie would "die." During the period 1948–1962, nitrates in Lake Erie increased 50%, while phosphate increased over 400%.[44]

The soil under long-term livestock feedlots has been found to contain 2000–4000 pounds of nitrate nitrogen per acre.[45] High nitrates in groundwater and drinking water is a common phenomenon in agricultural states, and now pesticides are a common pollutant of groundwater. In 1980, a Wisconsin study found 311 of the state's 11,396 small noncommunity water systems had more than the allowable concentration of nitrates, and a 1982 study found synthetic chemicals in 51 of the 208 community wells tested.[46]

". . . once your water *is* polluted, it could stay that way the rest of your lifetime, your children's lifetime, maybe even your grandchildren's lifetime. Groundwater is our most fragile resource: we don't have the technology to clean it up once we mess it up."[47] Since groundwater is found moving slowly in tiny soil and rock crevices, contaminated groundwater is practically impossible to clean up. Stopping the source of pollution and allowing natural water being added to the groundwater to dilute the polluted water would take decades. Polluted water pumped to the surface can be purified, but at great added expense.

[43]D. Pimentel, p. 62–63 in *Pest Control Strategies*, 1978, P. de Jong, ed.

[44]J. Verduin, p. 163–164 in *Agriculture and the Quality of Our Environment*, 1967, N. C. Brady, ed.

[45]*Ibid.*, p. 180.

[46]*Wisconsin State Journal*, April 10, 1983.

The verdict. The evidence is clear—and is mounting daily—that unwise and excessive use of agricultural chemicals (plus the additional sources of industrial and urban chemicals and wastes) is horribly polluting and poisoning our environment—and us. It can't go on much longer!

Weeds too. Not only has spraying millions of pounds of insecticides failed to keep the bugs under control, but problems are increasing with herbicides and weeds. In a speech to the North Central Weed Control Conference in December 1980, G. F. Warren of Purdue University admitted, "We have seen the rapid increase in resistant weed species almost from the beginning of commercial application of selective herbicides.[48] Weed species which never were a problem in our grandfather's day have now invaded our prime agricultural land. Herbicides that used to kill weeds now are either ineffective or have to be used at increased rates. Pesticide expert George Ware of the University of Arizona says that our pesticide arsenal is not keeping up with our pest problems because of resistance. "This health-conscious generation knows the undesirable effects that chemicals in excess—not just pesticides, all chemicals—can have on our environment. And more and more our farmers are beginning to know what their cost-benefit ratio is. . ."[49]

Less variety. Another phenomenon of our modern agricultural system is the decreasing variety and number of crop plants. According to J. R. Harlan, "most of the food for mankind comes from a small number of crops and the total number is decreasing steadily."[50] Botanist P. C. Mangelsdorf lists 15 crops that feed the world: wheat, rice, corn (maize), sorghum, barley, beans, peanut, soybean, sugarcane, sugarbeet, sweet potato, manioc, potato, banana, and coconut.[51]

According to 1969 figures, there are 269 varieties of wheat, but only 9 varieties comprise 50% of the wheat acreage. Only 6 major varieties of corn (maize) account for 71% of the planted acres, even though there are at least 197 commercial inbred varieties. Only 4 out of 82 varieties of potato are planted on 72% of potato land. And so on for other major crops.[52]

Beginning in the late 1960's and early 1970's, many native varieties of crops were replaced by the high yielding hybrid varieties of the "green revolution," as well as hybrids in general. Also, increased development,

[47]Brochure: *The Well on Your Farm*, Publication 5-3200, Wisconsin Dept. of Natural Resources, 1982.

[48]*Proc. North Central Weed Control Conference*, 1980, vol. 35, p. 3.

[49]AP news release, April 6, 1983; G. Ware, *Pesticides: Theory and Application*, 1983.

[50]J. R. Harlan, *Crops and Man*, 1975, p. 80.

[51]*Ibid.*

[52]*Ibid.*, p. 257.

urbanization, and destruction of natural habitats all around the world has threatened many wild varieties of crop plants. The many varieties of the crop species are a reservoir of genetic traits, some of which could potentially provide valuable characteristics for future crop improvement, such as increased food value, or disease resistance, or drought tolerance.

If those crop varieties become extinct, the potentially useful traits they carry will be lost. Many crop scientists and botanists are worried about this "narrowed genetic base" and are establishing seed "banks" to preserve as many varieties as possible.[53]

Planting large acreage in only one or a few varieties can cause other problems, such as allowing rapid spread of pests or diseases to which the variety is susceptible. This is best illustrated by the disastrous wildfire spread of southern corn leaf blight in 1970, when 80% of the U. S. corn crop was planted in hybrid varieties susceptible to the disease.

Hybrids. Hybrid varieties are probably man's first "high tech" achievement in agriculture, although the earliest hybrids were made before the principles of genetics were understood. A hybrid is the offspring that results from crossing of two different strains or varieties. The hybrid contains a combination of genetic traits or characteristics from the two parents. By applying the laws of genetics, breeders can "engineer" or predetermine the characteristics of the hybrid, such as yield, height of plant, and resistance to a certain disease. These man-designed varieties are uniform for certain characteristics, which is supposed to be an advantage; thus all the plants grow to the same height, mature at the same time, and produce uniform grain.

But the trouble is just that—they are all alike for those traits—there is no variation such as there is in open pollinated (non-hybrid) varieties. The individual plants of open pollinated varieties have slightly differing characteristics, which gives the total population of plants more versatility in withstanding environmental stresses and producing a good yield under nearly any conditions. On the other hand, hybrids are so "fine-tuned" to produce well under a certain limited set of conditions (soil type, fertility, weather), that if conditions are unfavorable, the crop may be a disaster.

Furthermore, the particular genetic traits which the scientists "build in" to the hybrid are not always wisely chosen. Generally, the number one emphasis is YIELD; that is, bulk—bushels, tons. Much less attention is paid to QUALITY and nutritional value. Breaking yield records is much more important than growing high quality food that animals or

[53]*Science*, 1981, vol. 214, p. 161–164, 421–423; 1984, vol. 224, p. 251–256.

people can thrive on. Amino acid, mineral, and vitamin content are way down on the list.

Tests of hybrid corn varieties by Ernest M. Halblieb of McNabb, Illinois, revealed that the hybrids did not take up seven to nine trace minerals. None absorbed cobalt, a mineral needed by animals in vitamin B_{12}.[54] Protein content of wheat and corn has steadily declined during the 20th century.

The system. But that's part of the system—the establishment—the interconnected association of government, industry, and education that runs our society, agriculture included. Much evidence shows that in recent decades, the system is not being run to benefit the farmer. Government programs and policies are oriented toward providing the urban consumer with low-priced food ("cheap food policy") and interconnecting the world's exporting countries by trade commitments ("one worldism").

Market prices for farm commodities are manipulated to force farmers to increase production and become "more efficient" (larger farms, more mechanization, higher yield per unit cost). Interest rates and costs of machinery, fuel, and fertilizers are now extremely high, which forces the smaller, "inefficient" farmer out of business.

U. S. farm and economic policies are set by politicians, economists, bureaucrats, businessmen, and scholars—NOT FARMERS! The President's Council of Economic Advisers, the Committee for Economic Development, the Council on Wage and Price Stability, the Joint Economic Committee of Congress, and the Federal Reserve System are in charge—NOT FARMERS!

In a 1962 document titled "An Adaptive Program for Agriculture," the Committee for Economic Development, a prestigious group of 200 leading corporate businessmen and scholars, spelled out a long-range program to move people and land out of agriculture by discouraging young people from entering farming careers, providing urban jobs for farmers, and manipulating (lowering) the price support system to discourage farmers from farming. Their program is succeeding *very* well (see *The Rest of the Story*, Chapters 7 and 8, for more details).

So there you have some of the fruits of our "modern" agricultural system. Not very appetizing, are they? In fact, they are rotten!

[54]C. Walters, Jr. & C. J. Fenzau, *ibid.*, p. 38.

Our modern economic and agricultural system is on the verge of collapse—if not in the process of collapse!

There HAS to be a better way. There IS a better way, and it's coming soon. It's a way of farming that isn't based on greed. That treats the soil with respect. That gives back, not just takes.

Some amazing things are in store for the future. Let's take a look at the coming revolution in agriculture.

CHAPTER 2

Nature's Way is Best

"And God saw every thing that he had made, and, behold it was very good." Genesis 1:31

It seems to happen so often that you can almost depend on it: whenever man comes up with some great new scheme to improve on nature, or some fantastic new technology or product, it eventually boomerangs on us. It turns sour. It has unexpected and undesirable consequences.

Witness the building of the Aswan dam, blocking the annual natural rejuvenation of Egypt's farmlands, plus increasing problems from parasitic human blood flukes. Witness the development of nuclear power generating plants, now being closed down as uneconomical and dangerous. Witness the massive flooding resulting from clear-cutting forests and farming steep hillsides the wrong way. And of course witness the horrible pollution and increased pest problems after four decades of using ultra-toxic herbicides and insecticides, covered in the previous chapter.

Natural laws. Nature operates under certain laws. Laws of physics and chemistry determine properties of matter and how chemical elements combine. Biological laws control how living species of plants and animals grow, reproduce, and interact. The science of *ecology* studies how different species interact with each other and with their environment. Ecologists know that the many species of plants and animals in a natural habitat (such as a forest or grassland) live in a system of checks and balances, sometimes called the balance of nature. Any particular species obtains food (stored energy) from certain sources (green plants make food with the aid of light energy; animals eat plants or other animals; bacteria and fungi live mostly on dead organic matter). One species keeps another in check. Soil is the basis for the growth of plants, and when plants and animals die, their bodies (plus animal wastes) are recycled into the soil to provide nutrients for plants. This beautifully designed system, or *ecosystem*, can be diagrammed in a simple way as follows:

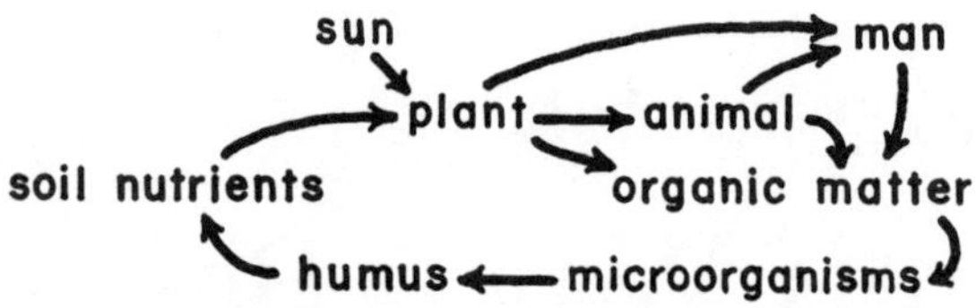

An ecosystem is similar to a self-regulating machine, made up of many different parts. When the machine is in good condition, well oiled, and with no loose bolts, it operates fine. Any excessive growth or reproduction by one species is soon controlled by its natural enemies. The farmer is a part of a modified ecosystem, since he grows plants and/or animals—living species. But he grows them under unnatural conditions: by themselves in a plowed field or in a barnyard rather than surrounded by many other species. Under these conditions, it is easier for things to get out of balance. If the crop plants or the animals experience stresses, then diseases and pests more readily attack, and when they do, cause more damage than they would in a natural ecosystem such as a forest or grassland. The fewer the species in an ecosystem, the more easily the balance of nature can be upset.

Effects of man. Also, in a "farm ecosystem," there is a much greater impact from man's activities than in a natural ecosystem. The farmer is constantly doing something—plowing, fertilizing, spraying pesticides, harvesting. These activities are intended to increase the crop's yield, but can create stresses for the crop if they are harmful to the ecosystem. Stresses on the crop plants lead to later problems—perhaps slowed growth, reduced yield, or pest attack.

As one simple example, cutting corn roots during careless cultivation (to kill weeds) will harm the corn plants' ability to absorb water and nutrients, thereby slowing growth, and allowing rootworms and stalk rot fungi to invade the damaged roots.

Soil is number one. As another example of how farming practices affect the "farm ecosystem," consider the soil, which is the all-important base from which our plant and animal food grows. *Mistreatment of the soil is the number one source of crop problems—and later, animal problems.* The average farmer simply doesn't realize how complex soil is nor how important good soil conditions are to growing healthy, high-yielding, high quality crops. Even bad weather conditions will not harm crops as much when soil conditions are good as they will when the soil is "sick." The laws of nature were designed to produce abundant, healthy, nutritious plant growth and healthy animals that eat the plants—provided the soil is "healthy." If one part of the ecosystem is out of

Ecology Simplified

In his book, *The Closing Circle* (1971), Dr. Barry Commoner cuts through the complexities of ecology and the problems of pollution, and exposes the causes of problems. He simplifies the principles of ecology into four laws:

1. "*Everything is connected to everything else*." That is, all parts of the ecosystem are interrelated, and a change in one part will affect others.
2. "*Everything must go somewhere*." Matter is not destroyed; it is merely transferred from place to place, converted from one molecular compound into another, and acts on the life processes of any organism it becomes part of.
3. "*Nature knows best*." Any major man-made change in a natural system is likely to be detrimental.
4. "*There is no such thing as a free lunch*." Every benefit is gained at some cost because of the interconnected ecosystem (law no. 1); payment of the cost cannot be avoided, only delayed.

Dr. Commoner traces the blame for our present ecological troubles first to the great increase in technology and production of pollutants after World War II (p. 140, 144), and second to a failure to recognize the importance of nature's complex systems and to see the overall effects of our actions (p. 189). But he also goes a step further toward the ultimate cause in pointing out the motive of profit-seeking (p. 258).

whack, the next part is thrown out of balance—and the next—and the next. Problems develop. Diseases and pests attack. Everything goes wrong. Let's take a closer look at the all-important *soil* and see how it functions—or should function.

Soil parts. Soil is not "just dirt." It is a very complex substance composed of many parts and undergoing many activities. Typical soil is made up of four main parts:

1. *Minerals* (about 45%), which come from the underlying rocks. This part of the soil provides bulk and contains a source of certain plant nutrients (calcium, magnesium, phosphorus, potassium), although not all of the nutrient supply will be readily available to plants because most nutrients are "locked up" in the chemical molecules of the minerals. The smallest mineral particles, especially those said to be in the *colloidal* size range, less than 0.002 mm (1/5000 inch)(this includes the smallest clay particles), are important because they hold nutrient ions by electrical attraction forces.

2. *Water* (about 25%), needed as a large part of plant cells and to

support the life of soil organisms. Also, water dissolves and carries plant nutrients.

3. *Air* (about 25%), which provides oxygen to roots and soil organisms.

4. *Organic matter* (about 1–5%), which includes both the living soil organisms and the dead organic matter which decomposes into humus. Humus contains tiny colloidal particles and can store even more nutrients than colloidal mineral particles (about three times more). The humus and living organisms have other important functions, covered later.

Classifying soil into these four components doesn't begin to reveal the great complexity of soil and what actually goes on in healthy soil.

Soil could be compared to a boiling pot of alphabet soup. It contains many different kinds of solid chunks, surrounded by water with many substances dissolved in it, all in constant motion. That soup analogy is not quite complete, since there should be as much air as water, plus a whole host of squirming animals and simple plants (bacteria, fungi) swimming around in it. But at least you begin to get the idea that soil is not a uniform dead powder or mud; it is active, dynamic, constantly changing.

Ions. There are inorganic ions and molecules, some dissolved in the film of water clinging to soil particles, but most held (adsorbed) on the surfaces of the colloidal clay and humus particles. These inorganic ions include some with a positive electrical charge (called cations), such as magnesium (Mg^{++}), iron (Fe^{++}), ammonium (NH_4^+), and hydrogen (H^+). Other ions (called anions) have a negative electrical charge. They mostly occur in the soil's water, and include nitrate (NO_3^-), phosphate ($H_2PO_4^-$), sulfate (SO_4^{--}), and hydroxide (OH^-).

Colloids. The colloidal clay and humus particles, which have a HUGE surface area and a large number of electrically charged sites, play an extremely important role in soil fertility since they hold most of the nutrient ions. The more colloidal particles, especially humus, the more fertile the soil can be, provided other soil conditions are right.

pH. The relative balance of hydrogen (H^+) and hydroxide (OH^-) ions is measured by the pH scale (technically, the pH scale measures just the hydrogen ions). The pH scale measures acidity and alkalinity, since

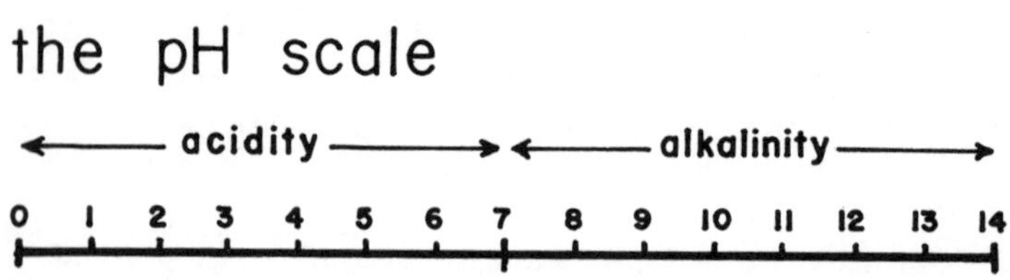

hydrogen ions cause acidity and hydroxide ions cause alkalinity. The pH scale runs from 0 (most acid) to 14 (most alkaline), with 7 being neutral. The pHs of some common materials are: lemon juice = 2, vinegar = 2.5, black coffee = 5, pure water = 7, sodium bicarbonate solution = 8.2, ammonia water = 11. Soil pH generally runs between 4 and 10, but most crops do best under slightly acid conditions (6.0–6.8).

Soil pH has little or no direct effect on plants, but too low or too high pH can be bad in various indirect ways. The various soil nutrient elements can become more or less available to the plant depending on pH. For example, the very important element phosphorus is most available from pH 6.5 to 7.5. Also, some elements can be so available that they become toxic; for example, manganese can be toxic to plants below pH 4.5.

Soils with an excess of salts and/or a high pH, such as are common in the western U. S., generally have poor soil structure (are poorly aerated), and thus limit crop growth.

Too high or too low pH can also limit the occurrence of beneficial soil organisms. Earthworms and soil bacteria are rare in very acid soils (below about pH 4.5), and the poorly aerated alkaline soils will be sparsely populated as well.

The trouble with soil pH is that the people who test soil and recommend fertilizers put the wrong emphasis on it. First of all, they seldom recommend that the farmer have his pH tested more often than once every two or three years. Actually, the soil pH is always changing, often by a factor of 200 to 400% during a growing season. So a pH reading taken on any one day every two or three years is pretty inadequate. The soil pH can vary depending on moisture, salt and ion concentration, carbon dioxide content, soil organism activity, and the depth from which the sample is taken. Also, the laboratory method used to measure soil pH can give differing results. If the soil is mixed with pure water, a higher reading will result than if a dilute solution of potassium chloride is used. Using an electronic pH meter, color indicator chemicals, and titration methods can all give different results. And on a microscopic scale, the actual pH among the soil particles can vary considerably, say at a root's surface compared to only a fraction of a millimeter away. The important thing is the pH immediately next to roots, which can change quickly and can be much different from the pH measured from bulk soil samples. Roots themselves can change the pH of their immediate environment.[1] Unfortunately, it is difficult to measure pH near the root.

[1]D. M. Huber & R. D. Watson, *Annual Review of Phytopathology*, 1974, vol. 12, p. 152; R. A. Olsen, R. B. Clark & J. H. Bennett, *American Scientist*, 1981, vol. 69, p. 378–384; D. D. Kaufman & L. E. Williams, *Phytopathology*, 1964, vol. 54, p. 139.

Yet in spite of all these uncertainties, the soil testing lab treats a single pH reading as if it were a fixed, super-accurate measurement and bases certain soil recommendations on it, mainly lime application.

As long as the soil is well aerated, has abundant humus and organisms, and is not toxic, soil pH can vary considerably without harming crops. Generally, the overall "best" pH range for most crops is between 6.0–6.8.

Life. Besides the inorganic parts of the soil, a healthy soil should be teeming with life. The living organisms of soil are so important that some authorities do not consider lifeless soil as true soil. A tiny lump of topsoil may contain millions or even billions of organisms.

The variety of soil organisms is large, ranging from microscopic plants (bacteria, actinomycetes, fungi, algae) to microscopic animals (protozoa, nematodes, mites) to larger burrowing animals (insects, earthworms). The numbers of these organisms in an acre of temperate grassland soil are shown in the following table.[2]

organism	*number/acre*	*lbs./acre*
bacteria	800,000,000,000,000,000	2600
actinomycetes	20,000,000,000,000,000	1300
fungi	200,000,000,000,000	2600
algae	4,000,000,000	90
protozoa	2,000,000,000,000	90
nematodes	80,000,000	45
earthworms	40,000	445
insects & other arthropods	8,160,000	830

The most abundant and most important soil organisms are bacteria, actinomycetes, and fungi. Bacteria are single-celled organisms that can live in either aerobic (aerated, with oxygen) or anaerobic (without oxygen) conditions, but generally cannot live in very acid soils. The aerobic bacteria are the most beneficial, while most anaerobic bacteria cause problems (some, however, help release inorganic mineral nutrients or fix nitrogen).

Actinomycetes are filamentous (thread-like) organisms that are considered to be intermediate between bacteria and fungi. They require aerobic, neutral or slightly alkaline soils.

Fungi, or molds and yeasts, are sometimes single-celled, but usually filamentous. They are mostly aerobic and can tolerate fairly acid conditions.

[2]L. M. Thompson & F. R. Troeh, *Soils and Soil Fertility*, 4th ed., 1978, p. 111.

These three types of microorganisms, with help from the other soil organisms, form a "volunteer army," performing an amazing variety of beneficial functions—if we just provide them with a comfortable home and some food. Just look at some of the things these marvellous creatures do for you.

Decomposition. Similar to the way our digestive system can break down the complex food substances we eat, a "living" soil is a "stomach." The variety of plants and animals in soil are like a chemical factory. They can take just about any substance and change it into something else. They can take the very resistant substance cellulose, a major component of plant cell walls, and break it down into simple sugars. Huge protein molecules are changed into "chunks" called polypeptides and later into amino acids, albumins, purines, pyridines, amines, ammonia, carbon dioxide, water, and miscellaneous ions. Even such highly toxic materials as man-made insecticides, herbicides, and industrial chemicals are attacked and broken down by certain soil microorganisms. In fact, in some parts of the U. S. corn belt, soil bacteria are gobbling up carbofuran insecticides and thiocarbamate herbicides as soon as they are sprayed on the fields—and the pesticides don't even have time to kill the weeds or insects they were supposed to! You can't outsmart "mother nature"!

Soil organisms also help break down and release nutrients stored (locked up) in the mineral particles of the soil, such as phosphorus, potassium, calcium, magnesium, and trace elements. Acids produced by soil microbes (including citric acid, oxalic acid, tartaric acid, malic acid, and galactouronic acid) dissolve and make available (chelate) mineral nutrients.[3]

The main service soil life performs in decomposing things is in recycling fresh organic matter (crop residues, animal manures, green manures) into that magical and mysterious substance, HUMUS. If it were not for this nutrient recycling, life on the earth would grind to a halt. We will say more about humus later.

In the natural ecological cycle, plants grow from the soil, animals eat part of the plants, and the dead plants, animals, and animal wastes "feed the soil," to be changed back into simple nutrients usable by a new generation of plants. It works beautifully without man meddling with it: forests and prairies grew for thousands of years. "Primitive" agriculture, not motivated by greed, imitates this natural cycle and grows nutritious food. It is only when greedy people who want to take more from the soil than they give back start short-circuiting the natural cycle that problems arise.

[3] *Soil Science*, 1952, vol. 72, p. 173–181; 1963, vol. 95, p. 105–114; *Annual Review of Plant Physiology*, 1968, vol. 19, p. 80–83; *Plant and Soil,* 1973, vol. 38, p. 191–201.

Roots

Beneath the soil's surface lies the dark, poorly-known world of your crop's roots. There are complex and ever-changing relationships among (a) roots, (b) soil particles, and (c) the living organisms of the soil, especially those that live directly on the root's surface.

Many plants have as much or more tissue below the ground (the root system) as they do above (stem, leaves). The total length of a plant's roots is astounding. In 1937, H. J. Dittmer measured the roots of a single rye plant. The main and smaller roots had a total length of over two million feet (387 miles) and a surface area of 2,554 square feet. The tiny root hairs attached to the smaller roots had a total length of 6,604 feet (1.25 miles) and a surface area of 4,321 square feet. Root hairs grow in intimate contact with soil particles and are the root's main absorptive area. Corn roots can easily reach down six feet in deep, loose soil.

The primary water and nutrient absorbing region of roots is near their tips, and a single plant can have thousands or millions of "feeder" root tips. Throughout most of a plant's life, new roots are constantly growing out to explore new areas of soil for water and nutrients. At the same time, many older roots are dying off. This death of roots (plus the entire root system when the plant dies) adds tremendous amounts of organic matter to the soil, plus aerates and loosens soil. Plants have about twice as many roots as they need to survive.

Roots are actively alive. They consume oxygen, give off carbon dioxide, and release secretions (called root exudates). When the carbon dioxide contacts soil moisture, it changes into a weak acid, carbonic acid. This acid, plus others from soil microbes and humus, helps break down soil mineral particles and releases (makes available) locked-up elements such as phosphorus, potassium, magnesium, calcium, and trace elements. It has been found that if a plant becomes short of such an element, it causes a sudden increase in acidity immediately next to its roots (pH can drop from 7 to 4, an extreme change), causing more nutrients to be released from the soil, satisfying its malnutrition.[4]

The root exudates contain sugars, amino acids, organic acids, enzymes, vitamins, and hormones. They help to nourish the rhizosphere organisms, special bacteria, actinomycetes, and a few fungi that live and proliferate on the root's surface. Most of the rhizosphere organisms are beneficial, making soil nutrients more available, helping the plant absorb nutrients, giving off growth-stimulating substances, and protecting from root diseases. Included among the root organisms are the mycorrhizae, symbiotic fungi that live partly inside the roots of most kinds of plants. They assist in the above-mentioned activities and can improve the health and yield of a crop.[5]

[4]R. A. Olsen, *et al.*, *ibid.*

[5]R. S. Russell, *Plant Root Systems*, 1977, p. 113–135; *Phytopathology,* 1978, vol. 68, p. 1377–1383; 1980, vol. 70, p. 1078–1082; 1981, vol. 71, p. 642–644.

Nourishing plants. Ways of "feeding the plant" (soluble fertilizers) rather than the soil give larger yields. But this depletes soil fertility and produces hard, compact soil—plants are then stressed, develop diseases, and are attacked by pests. The high level of soluble nutrients causes plants to "force feed," often leading to poor quality, nutritionally out-of-balance food for animals or people. Never mind, we spray on poisons to kill the pests and develop "improved" hybrids that give still larger yields while resisting the diseases. Of course, the low quality food and pesticide residues contribute to worsening health problems in livestock and people. But that's OK, we have lot's of talented vets and doctors, their black bags bulging with the latest drugs. Pretty stupid, isn't it? Trying to circumvent the natural laws causes problems. Problems cause more problems.

How else do soil organisms help us? After they have broken down organic matter to replenish the soil's supply of plant nutrients, some soil organisms actually help feed the nutrients to the plant roots. The space immediately surrounding a root (or root hair) is a different world than most of the soil. It is called the *rhizosphere*. Compared to elsewhere in the soil, the rhizosphere is a beehive of activity. Bacteria and fungi are much more abundant in the rhizosphere (two to ten times more), and they are of different species than elsewhere. These special rhizosphere organisms make nutrients more available to plants, sometimes by chemically releasing them from soil particles, or else by changing them into chelated forms.*[6] Special symbiotic fungi called *mycorrhizae* live partly inside the roots of most plants. Their thread-like cells explore soil particles, absorb certain nutrients (especially phosphorus, nitrogen, zinc, copper, molybdenum, and ammonium), and channel nutrients directly into roots. They also absorb more water than the roots alone could, aiding plants to survive droughts (see *The Rest of the Story*, p. 24–25).[7] They help plants tolerate high aluminum, manganese, and salts, which can be toxic. Nitrogen-fixing bacteria, actinomycetes, and algae increase the soil's supply of nitrogen—for free! But only if soil conditions favor their growth. Well-aerated, toxin-free soil promotes most species, although one type of nitrogen-fixing bacteria, *Clostridium*, is anaerobic. In one study of different sources of fertilizer and their effect on soybean growth, application of 5 - 10 - 10 NPK chemical fertilizer (at 980 lbs./acre or 49 lbs./acre soluble N) resulted in almost no root nodule develop-

• Chelates are organic molecules that can temporarily hold and release an inorganic ion.

[6]D. A. Barber, *Annual Review of Plant Physiology*, 1968, vol. 19, p. 71–88; R. S. Russell, *ibid.*, p. 113–140.

[7]*Acres U.S.A.*, December 1983, p. 4.

ment, while cattle manure and compost (31 tons/acre) resulted in well-developed, functional nodules (control plants, which received no fertilizer, had about 70% of their nodules functional). The beneficial nitrogen-fixing bacteria apparently came from the manure and compost, while the synthetic fertilizer inhibited them.[8]

Better health. Not only do rhizosphere organisms help feed plants, they also help keep the plants healthy. For one thing, some rhizosphere organisms release vitamins, hormones, and other secretions that increase the plant's growth. Vigorous plants resist pests and diseases. Plant roots also release secretions (carbohydrates, amino acids, growth substances) that favor the microorganisms' growth—a fair trade.

Also, some common rhizosphere organisms release antibiotics that kill or inhibit disease-producing microbes, such as root rot fungi. Others release materials called siderophores, which use up iron that is needed by the disease microbes, which then sort-of develop a case of "anemia" and die out. Recent recognition by plant pathologists of the ability of disease-suppressing soil organisms to markedly increase crop yields has stimulated research along these lines. Without diseases, strawberry yields can increase from 5 or 6 to 19 or 22 tons per acre![9] Recent research at the University of Illinois has found that inoculating soybean seed with certain strains of natural soil bacteria can protect plants from phytophthora root rot in stressful conditions. Yields were no different from plants protected by a fungicide or from phytophthora-resistant varieties.[10] Experiments at the University of California-Berkeley found that beneficial rhizosphere bacteria increased the yields of sugarbeets up to 15%, potatoes 33%, and radishes 144%.[11]

Some soil fungi are designed to trap and eat nematode worms, some of which invade plant roots. These fungi produce tiny donut-like "nooses" or sticky knobs and branches. When a nematode pokes its body through a noose or touches a sticky knob, it is immediately held tight, to become the fungus's dinner.

High but balanced soil fertility has long been known to increase the growth of rhizosphere organisms and to increase a plant's resistance to diseases and pests. Adding calcium (lime), organic matter, and carbohydrates (sugar, molasses) also stimulates soil bacteria and improves plant health.

Better tilth. Soil microorganisms, as well as plant roots, secrete slimy by-products composed of polysaccharides (complex sugars). They are

[8]W. F. Brinton, Jr., p. 109–117 in *Basic Technics in Ecological Farming*, 1982, S. Hill & P. Ott, eds.

[9]M. N. Schroth & J. G. Hancock, *Science*, 1982, vol. 216, p. 1376–1381.

[10]*Farm Journal*, August 1982, p. D–8.

[11]*Acres U. S. A.*, July 1984, p. 15.

sometimes called slimes, gums, or mucilages. The slimes protect the cells that produce them, but they also have the important function of "gluing" together soil particles. This forms small "crumbs" or aggregates, or produces "good tilth." Good tilth or crumb structure is vital for healthy soil and plants. By being clumped into crumbs, soil thus has a good porosity or pore space. The spaces between soil crumbs allow air and water to easily penetrate. Surface crusting is decreased, and a subsurface hardpan will not form. Erosion by wind and water are not a problem.[12] Rain soaks in and is available in dry periods; less runoff means no devastating flooding. Good aeration keeps roots and beneficial organisms healthy (see Chapter 3). Soil is loose and easy to till. Roots grow better and deeper.

Just think of all those benefits from millions of microscopic organisms and a "glue" that they produce. Why would anyone want to do things to the soil that would cripple this volunteer army?

Temperature. The beautiful thing about this "soil-care system" is that it is self-regulating. The soil organisms will supply plants with just the right amounts of the right kinds of nutrients in the right balance—IF everything is working right. If soil moisture, aeration, and pH are OK. If adequate nutrients are there to draw on. If these things are suitable, then temperature is the regulating factor. The growth activities of soil organisms, as well as crop plants, increase as environmental temperature increases.[13] An old rule (Van't Hoff's law) says that growth and metabolic rates approximately double for each 10°C (18°F) rise in temperature, within a certain range of normal activity (about freezing to about 90°F). What this means is that the faster a plant grows and the more nutrients it needs, the more nutrients the soil life will provide. What a fantastic system!

Humus. It is impossible to separate the benefits of soil organisms and the humus they produce. They are inseparable—literally. The dead decayed organic matter—humus—is literally filled with the squirming, swarming organisms that produced it.

As we have mentioned, humus is composed of very small (colloidal) particles. This property allows it to hold tremendous amounts of materials—nutrients as well as toxins (humus is a valuable detoxifying agent in the soil). Humus is a complex mixture of dozens of kinds of organic substances: proteins, lignin, cellulose, hemicellulose, fats, carbohydrates, vitamins, hormones, organic acids, and other molecules. These came from the complex components that made up the cells of the plant and animal organic matter, plus the cells and secretions of the soil

[12] L. M. Thompson & F. R. Troeh, *ibid.*, p. 55, 58, 457–458.
[13] *Ibid.*, p. 74–75.

organisms that produced the humus.

Humus thus forms a storehouse of a great variety of plant nutrients and growth-stimulating substances. It buffers soil from drastic changes in pH, and protects from high salt concentrations and toxins. It aids in soaking up and holding water, and in producing good tilth. Thus it helps prevent erosion and makes soil easier to till. It helps provide favorable soil temperatures. In the early spring, its dark color absorbs heat and warms up the soil, while in the hot summer its high specific heat (ability to hold heat) evens out soil temperatures between day and night.[14]

In tests of composted cattle feed-lot manure, as little as 1 ton/acre, incorporated into only the top 2–3 inches of soil, greatly improved the absorption and retention of moisture and improved tilth *as deep as 12 inches*, sometimes in only a few weeks. Some of the compost's active ingredients can move rapidly into the soil.[15]

If humus was sold (via ads in farm magazines and on TV and billboards) for $500 a ton, and was promoted by the universities and fertilizer dealers as the hottest new fertilizer which also reduces erosion and pests —farmers would fall all over themselves rushing to buy it! But there's no profit in something anyone can make for themselves—for practically nothing—even if it is much better than the commercial products. It's sad, but that's how our system works.

> "Government and private research and advisory services have emphasized the development and use of chemical fertilizers and chemical pest control. Alternatives, such as compost and manure for fertilizer, biological nitrogen fixation, and biological control of pests, have suffered from comparative neglect. The main reasons for this neglect are the comparative lack of private sales potential for natural methods, and various institutional factors and prejudices that have discouraged their use . . . Most farmers are uninformed regarding alternatives to agricultural chemicals . . . Government neglect appears to be a result of a combination of special interest pressure, research difficulties, and the common tendency to underinvest in public goods."[16]

Where's the proof? All right, you say. So it's good to have some humus and "bugs" in the soil. But do they *really* make that much difference in my crops? Won't my yields drop drastically if I don't pour on the fertilizer and spray for the pests that always seem to come? Is there any *proof*

[14]R. D. Hodges, The Case for Biological Agriculture, in *Proc. Intern. Conf. on Granular Fertilizers & Their Production, 1977.*

[15]*F. Sims, p. 118–121 in Basic Technics in Ecological Farming*, 1982, S. Hill & P. Ott, eds.

[16]R. C. Oelhaf, *Organic Agriculture*, 1978, p. 226.

that natural methods produce higher quality crops and healthier animals? I'm not going to change to something that might bankrupt me!

Those are understandable concerns and fair questions. Years of practical experience and considerable research have shown that natural methods do work. They give good yields of highly nutritious food which contributes to healthy animals and people. Like any farming system, the success of any one farmer during any one year depends on a number of factors: weather, management skills and proper timing of operations, crops and varieties, soil type, etc. Just as the latest hybrid variety or herbicide can be a failure, so can natural farming methods if the farmer doesn't know what he is doing.

Yields. The most important thing to the farmer who needs to sell his crops to pay the bills is *yield*—quantity. Bushels, tons. Here the story is mixed. If a farmer uses strictly "organic" methods (no manufactured fertilizers or pesticides), then yields for some crops may fall somewhat below yields with conventional ("chemical") farming methods. In a careful comparison of 16 pairs of "organic" vs. "chemical" farms in the corn belt, William Lockeretz and his associates at Washington University, St. Louis, Missouri, found that for the years 1974–75, organic methods yielded 13% less than conventional methods for corn, 16% less for wheat, and 5% less for oats. But yields for soybeans were equal and hay yielded 30% better on organic farms.

What does this tell us? First, that different crops respond differently to natural methods and to supplementary fertilizers. Corn especially and wheat give much greater yields with an extra boost of nitrogen. Soybeans can supply their own nitrogen by root nodules. Hay is hampered by a lot of synthetic fertilizers or pesticides. Second, if you *have* to get large yields of corn or wheat on large acreage, strictly organic methods aren't for you. You will need to at least use some extra fertilizer. But you can use kinds that favor soil life, you can build up your soil's humus level, and you can hold off on herbicides and insecticides if they aren't needed. Later chapters of this book will give some pointers.

Speaking of yields, I can't refrain from mentioning a soybean field I saw in southeastern Whiteside County, Illinois. It was during the blistering drought of 1983. On August 30, the plants were green and healthy, hard to wade through, about waist high. The stems were loaded with pods. Because of the drought, most pods contained only two beans, but the neighbor's plants were already brown—burned up. This field yielded nearly 60 bushels per acre, while the average for the area was less than 20. The farmer did not follow strictly organic methods, but he used fertilizers that are compatible with natural systems—and he spent only $28 per acre on them!

Several comparisons of yields of "organic" and conventional farms in Europe found that for most crops, organic farms had yields equal to or slightly higher than conventional farms. Yields were lower for orchard fruit and in some cases potatoes, corn, and dairying. But in other cases, potatoes yielded six times greater than average conventional farms (in Canada), and dairy cattle produced 11% more milk per cow on less feed and for a greater number of lactations than with conventional methods (in Switzerland).[17]

Most research studies on natural farming methods have been done in foreign countries, such as the Soviet Union (yield increases of 10–25% or more have resulted from soil inoculation with bacteria), Israel, and European countries. Sad to say, such research has been scarce in the U. S., partly because of pressure (backed by grants) from the chemical industry to test their products, by being in a "research rut" testing conventional methods, and because of the fact that natural systems are so complex and hard to analyze. Some of the U. S. research to supposedly test natural methods and products has even been biased and poorly designed.[18] For example, in a study at the University of Illinois to compare conventional methods and no use of herbicides, standard non-toxic weed control (cultivation) was not used, so it is no wonder that test plots became overgrown with weeds and yields were low.[19] In comparisons of certain commercial organic soil amendments at the University of Minnesota, conventional test plots received herbicides, while apparently no weed control was done on the organic plots.[20]

Food value. Now let's look at the other side of the coin—the nutritional value of the crop. This is especially important if you feed your crops to your animals or are trying to grow high quality food for specialty markets, such as poultry feed and food for human consumption.

Food value is difficult to measure. We can try to measure protein content, vitamin content, or mineral content. Those are expensive to test for. Good food should have all of them and more—and in a certain proportion. Certain amino acids are essential, and in the right balance Generally, lab tests only measure one or two aspects, such as crude protein or acid detergent fiber. Such tests are only rough guesses of nutritional value (crude protein doesn't even measure protein; the nitrogen content is merely multiplied by 6.25, assuming that most proteins con-

[17]R. D. Hodges, *ibd.*

[18]R. C. Oelhaf, *ibid.*, p. 166.

[19]W. A. Conterio, D. Bateman & L. Christen, *Pollution Solution Plots*, Univ. of Illinois Cooperative Extension Service, 1971.

[20]G. D. Holcomb, S. D. Evans, W. W. Nelson & C. J. Overdahl, Soils Fact Sheets no. 22 and 23, Univ. of Minnesota Agricultural Extension Service, 1976.

tain 16% nitrogen; non-protein forms of nitrogen such as nitrates will give a false reading). A good analysis of food value would include an amino acid analysis and tests for vitamins and minerals.

In the area of nutritional value and food quality, natural farming methods win hands down over "chemical" methods. Use of purified manufactured fertilizers can lead to trace element deficiencies because those elements are missing from the purified products. Use of rock fertilizers and organic matter will usually supply all that are needed. Excessive use of nitrogen fertilizer produces extra crop growth (high yield) and even high "protein" (crude protein), but the protein is of lower quality and hard to digest. What happens is that high nitrogen produces lush plant growth of watery tissues. Its *dry weight* is actually lower than for crops grown under balanced fertilization. [21] Extra nitrogen accumulates as toxic nitrates and free amino acids. [22] Other results of high nitrogen fertilization include loss of flavor and keeping quality, increase in crop disease susceptibility, and decrease of vitamins and minerals.[23]

Over-use of other fertilizer elements, or imbalance, also causes problems. Too much potassium causes extra crop growth, but it is high in carbohydrates and cellulose (and thus lower in protein and minerals); the livestock disease grass tetany (hypomagnesemia) can occur because high potassium leads to low magnesium in crops. Low calcium leads to animal bone disease (osteomalacia).[24]

Many farmers using biological methods notice that their animals eat less feed, but do not fall in production. A number of scientific studies have shown that livestock require up to 15% less of biologically-produced feed.[25]

	oats (Clinton variety)			**wheat** (Thorne variety)		
	organic	chemical	%diff.	organic	chemical	%diff.
thiamine (B_1)	610 mcg	317 mcg	92	941 mcg	451.7 mcg	108
riboflavin (B_2)	152 mcg	56 mcg	171	278.2 mcg	120 mcg	131
nicotinic acid	19 mg	9.5 mg	100	89.6 mg	54.8 mg	63
protein	16.4 g	12.8 g	28	12.8 g	11 g	16
calcium	58.3 mg	46.7 mg	25	40.7 mg	33.4 mg	29
phosphorus	362.5 mg	373.4 mg	-3	374 mg	377.7 mg	-1

(g = gram, mg = milligram, mcg = microgram)

[21]R. D. Hodges, *ibid.*

[22]*Journal of Nutrition*, 1952, vol. 48, p. 461–476; 1953, vol. 51, p. 623–635; *Cereal Chemistry*, 1977, vol. 54, p. 1034–1042.

[23]*Qual. Plant Mater. Veg.*, 1970, vol. 21, nos. 1–2, p. 35–68; 1972, vol. 21, no. 4, p. 381–398; *Food Research*, 1958, vol. 23, p. 648–655.

[24]P. Syltie, *Acres U. S. A.*, September 1981, p. 10–11.

[25]R. D. Hodges, *ibid.*

Above are some measures of vitamin and mineral content of two crops grown with "organic" and "chemical" methods in 1951.[26] I can't say that these test plots were grown under ideal conditions, but at least these figures definitely show a great improvement in food value when crops are grown with natural methods. There are other "fine tuning" methods which we will describe in later chapters that can greatly increase yields and quality beyond what the above figures show.

Animal and human health. The best measure of the nutritional value of a crop is its effect on the animals (or people) that eat it. Does it help them to have robust health, rapid growth, successful reproduction, and high production of meat, milk, eggs, or wool?

We (and our animals) are today afflicted by an increasing number of health problems. Cancer. Diabetes. Heart disease. Brucellosis (undulant fever) in livestock. Difficult breeding, spontaneous abortions, and birth defects. Scours in young animals. Most diseases are said to be "caused" by some kind of "germ" (pathogen). You call out the vet and he pulls a magic medicine out of his bag. But more and more evidence is piling up indicating that diseases have their *ultimate cause* in faulty diet. A deficiency of this or that trace mineral or vitamin. Or an excess of some toxin. Something is out of balance. This puts stress on the animal's (or person's) body systems. Other environmental stresses may be added, such as toxic pesticide residues, poor ventilation, overcrowding, or lack of proper care and concern ("TLC," tender loving care). Eventually the body's natural resistance to disease (the immune system) is "worn down" and cancer develops, or germs invade.

Tests where human patients' diets were supplemented with plenty of trace minerals, including manganese, cobalt, copper, magnesium, and zinc, have shown that even diseases considered bacterial in origin are actually brought on by dietary deficiency. These included undulant fever (Brucellosis) and Malta fever.[27]

Dr. Ira Allison also found that cattle with Brucellosis had trace element deficiencies in their blood which healthy cattle did not have. Cattle fed trace elements were immune to the disease, even though they were next to sick ones. Another well-known example is grass tetany, which is caused by a magnesium deficiency.

It seems pretty obvious that the best way to fight health problems is to *prevent* them. Start with healthy animals and keep them that way. No need for veterinarian bills and dead animals. The animals need good

[26]J. I. Rodale, *The Complete Book of Composting*, 1970, p. 607.
[27]W. Albrecht, J. Allison & F. Pottenger, *The Merck Report*, 1949.

management: comfortable quarters, good water, and proper care. But most important, they need a proper diet. One with the right amounts and balance of nutrients—amino acids, vitamins, minerals—dozens of different nutrients. It's pretty complicated. You can buy expensive feed supplement mixes. Some work pretty well and others actually make your animals sick.

The simplest way to feed your animals a balanced diet is to grow your own high quality feed. Crops that grow in fertile, well-aerated soil are going to have high levels and a proper balance of nutrients. And they will automatically produce good health in animals that eat them.

In a series of studies in Germany, Dr. E. E. Pfeiffer compared white leghorn chickens raised on feed grown with well-rotted manure or with a mixture of inorganic mineral fertilizers (kainite, basic slag, calcium nitrate, potassium nitrate, and ammonia). Over a total of 16 months (2 year's testing), 10 hens on the manure program laid 3129 eggs, while 10 hens on the mineral program only laid 2472 eggs, a 26.6% difference.[28] He kept some of the eggs. After two months, 14% of the eggs from the mineral program had spoiled, while only 4% from the manure program were not good.[29]

[28]*Acres U. S. A.*, November 1981, p. 30–31.
[29]E. Pfeiffer, *Bio-dynamic Farming and Gardening*, 1943, p. 174, 176.

The beneficial effects of a good diet can often be seen in an animal's general appearance: the fur is smooth and sleek, and the eyes are clear, bright, and lively. There is no coughing or mucus discharge from the nose.

In another study in Europe in 1978, E. Aenelt and J. Hahn found that mice, rabbits, and bulls fed on "chemically fertilized" feed had lowered fertility: atrophied uteri and testes, fewer live births, lower sperm mobility, and higher biochemical stress than those raised on compost-fertilized crops. *Yet conventional feed nutritional analysis showed no difference between the two feeding programs!*[30] It just shows you the value of standard lab tests; more extensive lab tests (such as amino acid or vitamin analysis) probably would have shown a difference.

Too much nitrogen ("protein") in dairy cattle feed leads to ketosis (acetonemia, like low blood sugar in humans), scours, mastitis, and decreased milk production or growth. But too little nitrogen can cause retarded growth, poor milk conversion, and irregular heat periods.[31] A right balance is needed.

Poor quality, moldy feed, which contains highly poisonous mycotoxins, such as aflatoxin, can lead to worse cases of scours, severe anemia from a stressed liver, and gastroenteritis. Molds can impair an animal's immune system (white blood cells) and interfere with proper metabolism of vitamins A, D, B-complex, calcium, phosphorus, and protein.[32]

Natural vs. man-made. In spite of what the experts say, there are differences between synthetic vitamins, such as you might get in a commercial animal feed ration, and vitamins from natural sources (plants). Even though there may be no chemical difference in the molecule (which is what the experts are referring to), the synthetic vitamin is pure, with nothing else. Vitamins are known to work better in the body when in combination with other vitamins or minerals (for example, in humans, vitamin E is more effective along with vitamins A, B-complex, inositol, C, and the minerals manganese and selenium). Vitamins from natural plant sources will automatically be along with other vitamins, minerals, enzymes, and coenzymes. Also, there are two chemical varieties of vitamins, mirror images of each other. Natural vitamins are always the "left-handed" form, and the body can readily use them. Synthetic vitamins are a mixture of both left and right-handed forms. Some people are even allergic to synthetic vitamins.[33] Nature's way is best!

[30]R. C. Oelhaf, *ibid.*, p. 44.
[31]*Acres U. S. A.*, *ibid.*, p. 14.
[32]*Acres U. S. A.*, Fair Issue 1976, p. 17; July 1981, p. 34–35.
[33]*Acres U. S. A.*, February 1982, p. 19.

Does it pay? All right, you say. So natural farming methods do give better crops and produce healthy animals. But can I make a profit at it? I've heard "organic" farmers can't make ends meet. Where can I sell high quality crops above the market price?

Let's look at some facts. Natural farming methods do have different levels of expense than conventional (chemical-intensive) methods. Obviously the natural farmer saves money on chemical fertilizers and pesticides, but he has to make more trips over the field to cultivate weeds and perhaps spread manure. Just what is the bottom line on costs?

In 1974–75, a group of researchers at the Center for the Biology of Natural Systems, Washington University, St. Louis, Missouri, did a study (published in 1976) comparing the profitability of 16 matched pairs of "organic" and conventional farms in the corn belt. All of the farms raised animals and had manure available. The operating field crop costs averaged out like this:[34]

Cost in dollars/acres	Fertilizers, soil amendments	Field operations	Manure spreading	Pesticides	Other (seed, drying, hauling)	TOTAL
Organic farms	6.17	11.92	1.37	0.36	11.16	30.95
Conventional farms	17.33	11.09	0.90	5.23	13.77	47.32

So the organic farms averaged over $16 per acre lower operating costs for field crops. But on these particular farms, yields of corn, wheat, and oats were lower than on conventional farms (soybean yields were the same, and hay yields were higher) (see figures under **Yields** earlier in this chapter). So with somewhat lower yields, and assuming that the crops were sold at market prices rather than being fed to the animals, the overall net income per acre was nearly the same for the organic farms as for conventional—about $133 per acre. These figures don't take into account that high quality crops can be sold for premium prices to the right buyer, nor that use of pesticides can lead to long-term health problems in animals and people. Also, the organic farms required about 20% more labor for corn and soybeans and 5% more for wheat and oats. No attempt was made to study the effects of feeding better quality feed to the animals. And finally, we should point out that yields do not have to be lower if the right methods are used (explained later in this book).

[34]W. Lockeretz, R. Klepper, B. Commoner, M. Gertler, S. Fast & D. O'Leary, *Organic and Conventional Crop Production in the Corn Belt: A Comparison of Economic Performance and Energy Use for Selected Farms*, Center for the Biology of Natural Systems, Washington Univ., St. Louis, Missouri, 1976.

The Kutztown Report

In a recently-completed 5-year study of a 320 acre crop/livestock farm in eastern Pennsylvania, the Rodale Research Center found that the low-input methods used produced yields comparable or better than conventional farms, and at a profit, too.

The Kutztown farm, located on hilly terrain, uses contour strip cropping and crop rotations to protect the soil. The annual erosion rate is about half as much as on a local conventional farm, and soil nutrient levels are adequate even though no major additions of synthetic fertilizer have been made since 1972. Manure, crop residues, and legume rotations supply fertilizer needs. No pesticides are used, and reduced rates of herbicide are used on corn and soybeans to reduce time-consuming cultivation during hay cutting. Crops grown are alfalfa and red clover hays, barley, oats, rye, wheat, corn (grain and silage), and soybeans. Crops, except for 2/3 of the hay, are fed to the animals: beef cattle, hogs, and chickens.

During drought years (1980, 1981), corn yielded 47 and 32% more than the county average, while in a wet season (1982), it was 5% greater (the overall 5-year average was 25% higher than county average). Hay and soybean yields were about 25% greater than county and state averages, while small grains were the same or slightly below.

The overall cost of production for all crops was 10% lower than on a comparable conventional farm, with the various crops ranging from equal to 30% less than conventional. Corn and small grain production costs were about equal to conventional farms, while hay was 18% less and soybeans 27% less. Labor requirements were about 10 to 30% higher than on a conventional farm, and these costs were not included in the production cost figures. Greatest savings in costs were in low fertilizer and pesticide expenses. (M. N. Culik, J. C. McAllister, M. C. Palada, and S. C. Rieger, 1983, *The Kutztown Farm Report*, 92 p.)

The Washington University study found that the conventional farms used almost three times more total energy than the organic farms (fuel plus energy used to manufacture fertilizers and pesticides)—18,400 Btu's per dollar compared to 6,800 Btu's per dollar for organic. This study was made just after commercial fertilizer prices shot up in the 1970's, after oil prices were raised. They are still high and going higher. Both fertilizers and pesticides are made from fossil fuel. As prices for these continue to climb, natural farming methods have more and more of a cost advantage. In fact, this is probably one main reason the chemical industry is now seriously researching biological methods and developing "biotechnology."

A recent six-year study (1978–83) by the University of Nebraska compared several systems, ranging from continuous corn with the full array of commercial fertilizers and pesticides to crop rotation with manure (8 tons/acre) for fertilizer. The "organic" system won hands down, earning a net income of $125.95 per acre compared to $94.58 for continuous corn. The per-acre costs for 1983 were $106.20 for continuous corn, while the "organic" rotation only cost $59.92 (or $71.52 if oat straw were sold rather than being recycled).[35]

Finding premium markets or feeding high quality food to your own animals, and the freedom from depleted soil and toxic chemicals makes natural farming methods so much better than conventional methods that there is no comparison.

Modern, conventional agriculture is in trouble! It's days are numbered. More and more farmers are seeing the problems first-hand. Things can't go on much longer. Before we look at better ways of farming, let's consider some problems in more detail—anaerobic soil, and weeds and pests.

[35] *The New Farm*, November/December 1984, p. 56–57; Univ. of Nebraska College of Agric. Dept. of Agric. Economics, staff paper no. 14, 1984.

CHAPTER 3

The Villain of Anaerobic Soil

"And thy heaven that is over thy head shall be brass, and the earth that is under thee shall be iron." Deuteronomy 28:23

When you see a small lake in a field two days after a rain, or a stunted field of corn with yellowish leaves—do you wonder why—what went wrong? Or do you just shrug your shoulders and think, "Oh, that's the way all the fields are this year"?

We need to do more questioning, probing, and searching. We need to be aware of problems and then find out what's causing them. What's *really* causing them. Only then can we solve the problem—by attacking the real cause or causes, not just trying to cover up the symptoms as so many "experts" are doing these days.

Soil air. *One of the worst and most widespread problems today is anaerobic soil.* Now before you get scared off by that technical term, let's explain. *An-aerobic* simply means "without air," or "without oxygen." The opposite term is *aerobic*, meaning "with air" or "with oxygen." When these terms are applied to soil, they refer to the relative amount of air or oxygen in the pore spaces, or to well-aerated or poorly aerated soil. In the previous chapter we mentioned that typical soil contains about 25% air. This applies to topsoil, and it is a rough average. The pore space between solid soil particles occupies about 50% of the soil's volume (in good loose soil), and the pore space can be filled with either air or water. Usually the amount of water fluctuates between 15 and 35% (and the same for air; when the percent of water is higher, the percent of air is lower).

Sandy soils have large pores between sand grains and are usually well aerated. Clays have much smaller spaces between the fine clay particles and can easily become "tight" and poorly aerated. A soil with good crumb structure (good tilth; with aggregates—see Chapter 2) will be well aerated.

In any soil, the upper layers of the soil will be aerobic and have more air than deeper soil. Soil air comes from the atmosphere, so the deeper you go in the soil the less air there will be, and anaerobic conditions prevail. Actually, on a microscopic scale, there is usually a mixture of aerobic and anaerobic conditions. The center of a soil crumb may be anaerobic, while a few thousanths of an inch away, near a pore, there may be plenty of oxygen.

Importance of air. Why is air or oxygen so important in the soil? There are a host of interconnected reasons, but there are two main things in the soil that need oxygen: plant roots and most of the beneficial soil organisms. Roots need oxygen to function properly and to live. More about that later.

The beneficial soil organisms, most types of bacteria, plus fungi, actinomycetes, algae, protozoa, earthworms, etc., need oxygen to survive. They are called aerobic organisms, or aerobes. Some bacteria can live and grow without oxygen; they are called anaerobic organisms, or anaerobes. Some talented bacteria can switch back and forth depending on their environment's oxygen supply; they are called facultative anaerobes. Some of the anaerobic bacteria are beneficial (some decompose minerals to release nutrients; one type fixes nitrogen), but most are harmful to plant growth and the soil's nutrient and humus content. So, the better the soil is aerated (with good structure or tilth), the better it is for growing crops (in rare cases of over-aeration, humus will be "burned up" too quickly). We covered the beneficial functions of soil organisms and humus in the previous chapter.

Signs. What are the signs of anaerobic soil? Soils that are "tight" and heavy are more apt to be anaerobic. Crusting and compaction are bad signs. Another common sign is standing water and waterlogging, which indicate tight soil (little pore space) or an underlying hardpan (or plowpan, claypan). Humus levels are probably low, and often the soil is practically sterile, without earthworms and other life. If you dig up a plant (corn is especially good) and shake off the soil, the depth to which small feeder roots extend is a good indication of how deep your soil's aerobic zone is.

These signs of trouble are all too common in fields across the country today. Let's look at the problems anaerobic soil can cause.

Anaerobic problems. The number and variety of problems that can result from anaerobic soil are amazing. Some affect the plant directly and immediately, while others are delayed or indirect. They are summarized in the accompanying diagram.

Anaerobic soil conditions first of all means that the oxygen content is low and the carbon dioxide level is high. Well-aerated soil may have an oxygen content only slightly lower than in the atmosphere (21%), but in soil with poor structure or because of farming practices that decrease

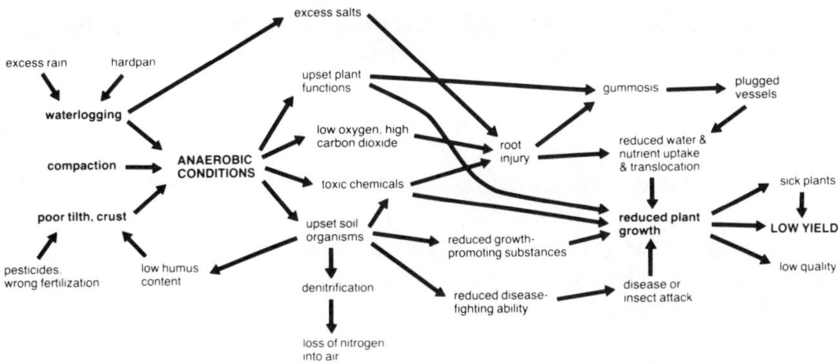

Problems of anaerobic soil.

aeration, the oxygen level may be below 1%.[1] Roots and soil organisms can deplete the available oxygen supply in a matter of hours.[2]

The atmosphere only has 0.03–0.04% carbon dioxide. Normal soil has about 0.1–1.5% (non-agricultural soil) to 3% (cropped soil). Anaerobic and waterlogged soils can have up to 15–20% carbon dioxide. These levels are "directly toxic to plants. . . the capacity of roots to absorb water and to a lesser extent, ions [nutrients], is sharply reduced."[3]

Changes in soil chemistry. These large shifts in oxygen and carbon dioxide content can cause adverse conditions in the soil. If the soil is high in calcium, high carbon dioxide forms carbonic acid, lowering pH (more acidity). A 10% carbon dioxide level decreases pH two units (100 times, since each pH unit is ten times the one above or below it). In low calcium soils, pH can be raised instead.[4] When pH is lowered especially, the availability of several elements changes and nutrient imbalances and toxicities can result. For example, aluminum and manganese become more available (and possibly toxic) at low pH, while nitrogen, phosphorus, potassium, magnesium, calcium, and sulfur become less available.

At low oxygen levels there are other chemical changes causing toxicities. Certain forms of iron and sulfur (ferrous and sulfide ions) can build up to toxic levels. Also the availability of several trace elements and heavy metals is increased, possibly to toxic levels (these include boron, cobalt, molybdenum, nickel, lead, and zinc).

[1]R. W. Pearson, p. 119 in *Plant Environment and Efficient Water Use*, 1966, W. H. Pierre, *et al.*, eds.

[2]B. D. van't Woudt & R. M. Hagan, *Agronomy Monographs*, 1957, vol. 7, p. 515.

[3]R. D. Durbin, p. 107–108 in *Water Deficits and Plant Growth*, Vol. V, 1978, T. T. Kozlowski, ed.

[4]R. D. Durbin, *ibid.*, p. 108.

Finally, under waterlogged conditions, certain anaerobic bacteria rapidly change nitrate nitrogen (which plants need) into nitrites (toxic to plants) and then into gaseous nitrogen or nitrogen oxides (which escape into the air). Close to 100 lbs./acre of nitrogen can be lost in a growing season due to denitrification.[5]

Furthermore, waterlogging dissolves excess salts, either from the soil's minerals or from previous fertilizer applications. Roots are injured, plant growth is slowed, and excess salts are carried to the leaves where they can accumulate and kill areas of the leaves of sensitive species ("leaf scorch"). The dying leaf or root tissues allow fungal diseases to get started.[6]

Upset soil life. In "healthy" soil, the many species of microscopic organisms live in a constantly fluctuating balance, with one species influencing another. Some help each other and some are "antagonists," keeping each other in check.

In waterlogged soils, "within a matter of hours the soil microflora changes dramatically, particularly in the rhizosphere. Fungi and actinomycetes almost disappear while many largely anaerobic bacterial species now flourish."[7]

The anaerobic bacteria carry on metabolic activities (feeding, excretion) sometimes called fermentation or putrification. They release a variety of chemical by-products which are toxic to roots in large enough amounts: hydrogen, ammonia, methane, hydrogen sulfide, carbon dioxide, and ethylene.[8]

Upset plant functions. Anaerobic conditions and waterlogging cause drastic changes in the functioning of the entire plant, perhaps only a few at first. But one problem leads to another, and pretty soon the plant is hopelessly sick, growth slows, and yields are cut. We have already mentioned how low oxygen, high carbon dioxide, and altered soil chemistry can disturb or damage roots. Only a short period of anaerobic conditions will retard a plant's functions, even though it will recover and appear to be normal and healthy. A single day of anaerobic conditions (produced by flooding) can reduce the dry weight growth of tomatoes as much as 30%.[9] Some crops such as soybeans, apples, and grapes are more resistant, however.[10]

[5]L. M. Thompson & F. R. Troeh, *Soils and Soil Fertility*, 4th ed., 1978, p. 248–249.
[6]R. D. Durbin, *ibid.*, p. 109.
[7]*Ibid.*, p. 110.
[8]*Ibid.*
[9]J. M. Fulton & A. E. Erickson, *Proceedings Soil Science Society of America*, 1964, vol. 28, p. 613.
[10]R. D. Durbin, *ibid.*, p. 103.

Hormone changes. At least three plant hormones can be thrown out of balance by anaerobic conditions. Ethylene production is increased, and the transport of cytokinins and gibberellins from the root is decreased. This causes reduced stem growth and reduced chlorophyll production in the leaves. Sometimes abnormal growth of roots from the lower stem and of "tumors" on the stem will occur. The disease called gummosis can be triggered (see next section).[11]

Gummosis. When a plant is under stress, it may produce slimy, sticky, gummy materials in its tissues. They can plug up some of the tubes of the vascular system, the vessels that carry water and nutrients upward from the roots (called the xylem tissue) or the cells that carry food from the leaves to other parts of the plant (called the phloem tissue). This condition of "plugged plumbing" is called gummosis by plant

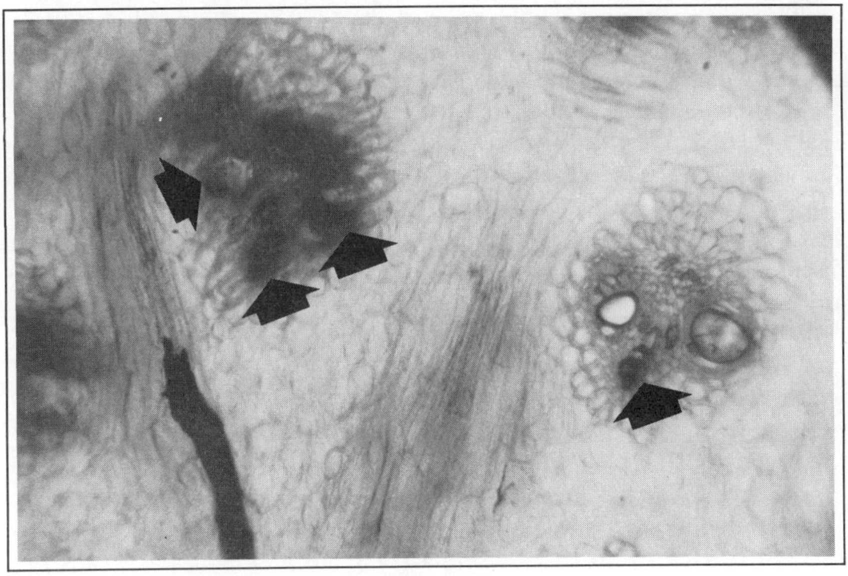

Vessels of corn plugged by gummosis. Magnified 100 times by microscope.

disease specialists. Leaf wilting, reduced growth, and death of some cells or of the whole plant can result.[12] Gummosis can be caused directly by anaerobic soil conditions or from toxins released by anaerobic bacteria, from unbalanced plant hormones, from disease or insect attack, and from soil nutrient deficiencies.

[11] *Ibid.*, p. 104–105.

[12] *Ibid.*, p. 110; P. W. Talboys, *Water Deficits in Vascular Disease*, 1968, p. 262; T. T. Kozlowski, *Water Supply and Leaf Shedding*, 1976, p. 214–215.

Fermentation. Because of oxygen deficiency in the soil, the plant's metabolism can be upset, with anaerobic respiration (fermentation) replacing the usual aerobic respiration. A by-product of fermentation is ethyl alcohol (as moonshiners and brewery workers know!). Alcohol levels in the plant's cells can build up to toxic levels.[13] In one study, it took only 2 1/2 hours for potato tubers to become anaerobic when covered by a film of water, which reduced the rate of oxygen uptake.[14]

Diseases and pests. Healthy plants can resist attack by disease pathogens ("germs") and pests (insects, mites, nematodes)(see Chapter 4). Stresses can cause a plant to lose its natural immunity. The stresses resulting from anaerobic soil can allow diseases and pests to get a foothold. Certain microorganisms which are normally present in soil and do not attack healthy plants are definitely known to do so to stressed plants under anaerobic conditions.[15]

Overview. Whew! That's quite a list of problems resulting from poorly aerated or waterlogged soil. In general, all of these results of anaerobic soil conditions have the effect of disrupting the soil's beneficial organisms and the normal functions of the plant. Thus the growth and health of the plant suffer and yields and/or crop quality are reduced. They are sources of unnecessary *stress* on the plant. Other factors can cause stresses also: drought, high temperatures, cloudy weather, unbalanced soil fertility, etc. Any stress is bad for the plant and can lead to diseases, pests, and reduced yield and quality. Why add unnecessary stresses? The harmful effects of anaerobic soil can be overcome *and prevented*. Let's see how.

The reasons. Before we can correct a problem, we must know what we're doing wrong. That way we can stop doing wrong things and start doing right things. Here are the main causes of anaerobic, tight, and waterlogged soils:

1. *Soil type and chemistry*. The physical and chemical make-up of the soil partly determine its texture and cohesiveness. The greater the percent of clay particles (as compared to silt and sand particles), the harder (when dry) or stickier (when wet) the soil. Greater organic matter (humus) content gives looser soil. Problems in the lower soil layers lead to hardpan, poor drainage, and waterlogging.

The particular chemical composition of the soil particles and the degree of electrical bonding of particles play a role in soil structure. Alkaline (high pH), saline (high salts), and high lime soils, such as are common in the western U. S. often have poor structure and aeration because

[13]J. M. Fulton & A. E. Erickson, *ibid.*
[14]W. G. Burton & M. J. Wigginton, *Potato Research*, 1970, vol. 13, p. 180–186.
[15]R. D. Durbin, *ibid.*, p. 109–110.

the particles do not form aggregates ("crumbs"). Their high pH and/or salt content also create other problems for agriculture.

2. *Low humus content.* As we have pointed out in Chapter 2, humus and the multitude of microscopic organisms that live in it (especially the bacteria) "glue" soil particles into aggregates and give it a loose, spongy structure. This increases pore space and aeration. On the other hand, low humus content and low populations of soil life lead to tight soil, crusting, and if in the lower soil layers, to hardpan and waterlogging.

3. *Traffic and tillage.* Frequent traffic, heavy machinery and working the soil when it is too wet lead to compaction and tight soil. Too much tillage of lifeless, low-humus soil makes it even worse because what organic matter that is left is oxidized ("burned up," destroyed) and there are few beneficial organisms to rebuild it.

4. *Wrong fertilization.* Some commercial fertilizers can alter soil structure directly, others indirectly. The worst is anhydrous ammonia. As liquid ammonia is injected into the soil, it changes into ammonia gas and has a number of harmful effects. The effects are most severe along the point of injection (a two to four inch diameter zone) and decrease considerably a few inches away.

Ammonia gas is a cell poison; therefore it can kill soil organisms and plant roots.[16] Studies in sandy soil found that ammonia caused a sharp decrease in numbers of bacteria, fungi, and nematodes in a 3-inch zone. Within a week to ten days, bacteria and nematode populations recovered, mainly bacteria that convert ammonia to nitrate and nematodes that feed on dead organisms. But even these kinds were inhibited for at least a month in areas of high ammonia concentration (over 300 parts per million).[17]

Also, there is a sudden rapid rise in pH (strongly alkaline) near the injection point. Depending on the soil type and moisture, the pH can jump to 8, 9, or 10, which is detrimental for crop growth. Gradually, the ammonia is converted to nitrate nitrogen by surviving microorganisms, and after several weeks the pH drops to a normal level. But then it may continue to drop, producing too acid conditions.[18]

There are sudden drastic changes in temperature when liquid anhydrous ammonia is injected into the soil. First, because of the immediate changing of the compressed liquid into a gas, there is an "instant freeze," with frost sometimes visible at the soil surface. Then, within a minute's

[16] M. M. Mortland, *Advances in Agronomy*, 1958, vol. X, p. 344–345.

[17] *Proc. Soil Science Society of America*, 1954, vol. 18, p. 178–181; 1955, vol. 19, p. 55–58.

[18] M. M. Mortland, *ibid.*, p. 342; R. I. Papendick & J. F. Parr, *Soil Science*, 1966, vol. 102, p. 199; J. F. Parr & F. E. Khasawneh, *Plant & Soil*, 1968, vol. 29, p. 466–467.

Applying anhydrous ammonia.

time, as the ammonia is being adsorbed onto the soil particles, a great deal of heat is generated from the chemical reactions involved. The temperature can shoot up as much as 20°F at 3/4 inch from the site of injection (in a silt loam). Within 15 minutes, the heat dissipates.[19] This torture surely can't do soil organisms any good.

Besides these short-term effects, anhydrous ammonia also causes undesirable long-term changes in the soil. It alters soil structure. The soil structure near the point of injection is changed so that the soil dries out more rapidly.[20] This could have disastrous consequences in a drought year. Anhydrous also "cements" soil particles (calcium, aluminum, and silica particles).[21]

Furthermore, ammonia makes humus much more water soluble, al-

[19]F. E. Khasawneh & J. F. Parr, *Soil Science*, 1968, vol. 104, p. 177–181.
[20]J. F. Parr & F. E. Khasawneh, *ibid.*, p. 460–470.
[21]W. Shuman, *Acres U. S. A.*, May 1981, p. 5.

lowing this valuable substance to leach away into the ground water.[22] To illustrate this, just drop a pinch of "leaf mold" or compost from your garden or flower bed into a glass of water. The humus will float on top and not dissolve. Now add a few drops of household ammonia, stir, and watch the dark color of dissolved humus spread throughout the glass. Now do you believe it?

Humus makes soil loose and well-aerated and holds water. If you remove humus and degrade the soil structure, *your soil will become harder*. This characteristic of anhydrous ammonia was used by the U. S. Army during World War II, to harden the ground on South Pacific islands to make emergency landing strips for bombers. They used high application rates, 700 to 2000 lbs./acre.[23] The anhydrous injection technology developed during World War II was then adapted to American agriculture, at lower application rates. That way it takes longer to turn your fields into an airport runway. As your soil becomes harder, you help your local equipment dealer stay in business as you buy a bigger tractor. That's progress!

The high rate of nitrogen supplied by anhydrous ammonia also causes other problems. Yields may increase, but crop quality goes down (less usable protein—see Chapter 2, **Food value**). Maturity is delayed; disease and pest resistance is lowered. After several years even yields may drop because the soil is so out of balance. Anhydrous ammonia may be the "cheapest" source of nitrogen, but it is no bargain!

If overused, urea and diammonium phosphate (DAP) can release large amounts of ammonia when they break down.

Another fertilizer element that may cause problems is *magnesium*. Plants do need a certain amount of magnesium, to manufacture chlorophyll and activate enzymes. Most soils have enough magnesium, although magnesium-deficient soils are found in parts of western Oregon, Washington, and west-central California, the Atlantic coastal plain, and the Gulf Coast; also some irrigated sandy soils and highly leached acid soils are deficient. Magnesium sulfate (epsom salts) and sulfate of potash-magnesia (sul-po-mag) can supply magnesium. Dolomitic limestone (dolomite, calcium·magnesium carbonate) can supply magnesium and is the cheapest source of lime in some areas. The trouble with dolomite is that it is a very hard mineral and takes many months to break down and release its calcium and magnesium.

But the problem with magnesium is that *in excess* it can cause nutri-

[22] J. F. Parr & F. E. Khasawneh, *ibid.*, p. 468; R. I. Papendick & J. F. Parr, *ibid.*, p. 198–199.

[23] W. Shuman, *ibid.*

ent imbalance and nitrogen loss, and harden soil. Excess magnesium can cause phosphorus and potassium deficiency. If calcium is low and magnesium high, organic matter can ferment and release alcohol, which kills soil life. Magnesium can "cement" clay particles (by increasing electrical bonding), forming a crust and anaerobic conditions.[24] Denitrification can then occur, with loss of nitrogen.

Other fertilizers that can lead to hard, anaerobic soil are the high salt fertilizers, which when used in too great amounts, kill soil organisms and eventually degrade soil structure.[25] The salt indexes of some common fertilizer materials are:[26]

ammonium nitrate	104.7
ammonium sulfate	69.0
calcium nitrate	52.5
calcium sulfate	8.1
diammonium phosphate	34.2
mono-ammonium phosphate	29.9
potassium chloride	116.3
potassium nitrate	73.6
potassium sulfate	46.1
sodium nitrate	100.0
sulfate of potash-magnesia	43.2
superphosphate	7.8
triple superphosphate	10.1
urea	75.4

The higher the salt index, the smaller the "doses" that should be used; also, keep high salt fertilizers away from the seed in the row or they may kill or stunt seedlings.

One commonly used fertilizer material, potassium chloride (muriate of potash, kalium potash) gives a double whammy. Not only does it have the highest salt index, but it is 47.6% chloride. Many crops are sensitive to chlorides (potatoes, tobacco, corn, soybeans, sweet potatoes, strawberries, cucumbers, peppers, citrus, and blueberries) if present at more than 100 parts per million. High chloride reduces sugar content and quality of crops.[27] Only 18 lbs./acre or less of muriate of potash are enough to injure seedlings. Soil bacteria are also very sensitive to chlo-

[24]T. L. Friar, *Acres U. S. A.*, Fair Issue 1976, p. 9; November 1983, p. 30; C. Walters, Jr. & C. J. Fenzau, *An Acres U. S. A. Primer*, 1979, p. 306; R. Wolf, *Organic Farming*, 1977, p. 154.

[25]E. W. Russell, *Soil Conditions and Plant Growth*, 9th ed., 1961, p. 435.

[26]*Farm Chemicals Handbook*, 1981, p. B74.

[27]*Jour. American Society of Agronomy*, 1919, vol. 11, p. 1–32.

rides. Anything that hampers soil life decreases humus production and can lead to anaerobic soil.

Finally, wrong fertilization with organic matter can cause anaerobic, toxic soil. Plowing under a lot of fresh organic matter (manure, green manure, by-product wastes), especially into moist soil, will cause a population explosion of soil microbes. Soon they will use up all the available oxygen. Anaerobic bacteria will take over and cause fermentation and the release of a host of toxic substances, such as hydrogen sulfide, methane, aldehydes, and alcohols. Also, excessive manure application greatly increases the soil's salt levels. Fresh organic matter should be worked into the upper, better aerated layers of the soil and not applied too heavily, especially in soil that is moist or tight. Compost (decomposed organic matter, essentially humus) has none of these disadvantages and can be added to soil in any amount.

5. *Pesticides.* Overuse of herbicides, fungicides, nematicides, and insecticides (in general, *pesticides*) often creates soil problems, including anaerobic, hard soil.

Some pesticides can change soil structure and kill or inhibit the soil microbes that produce humus. Some species of organisms may flourish, however. Thus the all-important *balance* of soil life is upset (more about this in Chapter 4). Humus level is gradually reduced and soil structure suffers. Erosion, crusting, and compaction increase. Anaerobic conditions result.

The herbicide atrazine is toxic to earthworms at 4 ppm (parts per million), but apparently does not kill most soil bacteria and actinomycetes at 70 ppm.[28] Some insecticides and most herbicides are not extremely toxic to soil microorganisms, and soil can be fairly quickly recolonized. However, anything that disturbs or even temporarily kills some of the soil organisms will upset the delicate balance that exists among the different species. Sometimes the kinds of microbes that recolonize pesticide-treated soil turn out to be harmful or pathogenic species.[29] Often there is a population explosion of species that are able to use the pesticide as food and break it down. This is basically good from the standpoint of getting rid of the poison, but again, the balance of species is upset.

The continuing use of herbicides year after year often promotes certain weed species, and some weeds are actually resistant to certain herbicides. In Hawaii, use of 2, 4-D on sugar cane led to infestations of foxtail and crabgrass, while the same herbicide used on wheat in Saskatchewan

[28] *Advances in Agronomy*, 1971, vol. 23, p. 212.
[29] J. P. Martin, *Residue Reviews*, 1963, vol. 4, p. 112.

resulted in three *new* weeds taking over: wild buckwheat, Canada thistle, and perennial sowthistle. Atrazine used on corn in Michigan brought good crops of fall panicum, green foxtail, crabgrass, and witchgrass.[30] Many of those weeds grow best in anaerobic soil (see Chapter 4).

Correcting anaerobic soil. Now that we have seen what causes anaerobic soil, we can then look at the other side of the coin and see how to correct problem soil and how to prevent soil problems. These corrective methods solve a multitude of problems, and they will be covered in more detail in Chapter 5.

1. *Increase humus and soil life.* If soil is not in extremely poor condition, work fresh organic matter (manure, crop residues, green manure, sewage sludge or other by-products) into the upper several inches of soil. This should be done in the fall if possible, or else a few weeks before planting a crop. The mixing of both plant and animal wastes gives better results, and the mixing of rock fertilizers (lime and soft rock phosphate; greensand if potassium is needed) gives even better results. Do not work fresh organic matter deeper in soil than the present aerobic zone or a little deeper (look at the depth of plant feeder roots to tell how deep the aerobic zone is).

If soil is very poor, work in compost rather than fresh organic matter. A little poultry manure added will provide an abundance of beneficial bacteria. Rock fertilizers mixed in will increase the benefit.

Leaving a field in a crop of a legume (clover, alfalfa, vetch, etc.) or especially perennial grasses (smooth brome, fescue, bluegrass, timothy, reed canary grass, etc.) for a year or more is one of the best ways to increase humus and improve soil structure (because of all the small roots).

Any of those things, cover crops, fresh organic matter, compost, and rock fertilizers (high calcium lime and soft rock phosphate especially) will help loosen tight soil, prevent crusting, reduce erosion, and increase water retention. A good earthworm population will break up a hardpan.

2. *Use temporary helps.* Until soil tilth improves, cultivate to aerate upper soil layers (cultivation is also important for weed control). Subsoiling will break up a hardpan, at least temporarily. Plowing a little deeper each year will gradually eliminate a hardpan, especially if humus is increased at the same time. Some commercial soil conditioners (wetting agents or surfactants, humic substances or humates) can help loosen soil and break up a hardpan, at least sometimes. They can be expensive, but may be a temporary help. Products that inoculate soil with algae or bacteria sometimes work well, but again may be expensive. Ground-up

[30]A. W. A. Brown, *Ecology of Pesticides*, 1978, p. 323–324.

rock fertilizers (lime, rock phosphate) generally help improve soil structure, and seaweed products can also have beneficial effects. See Chapter 7 for more details.

3. *Reduce compaction.* Use lighter machinery, avoid unnecessary trips over the field, and do not work soil that is too wet.

4. *Reduce tilth-destroying chemicals.* Reduce or eliminate the previously-mentioned fertilizers and pesticides that tend to destroy soil structure, humus, and/or life.

CHAPTER 4

Diseases, Weeds and Pests— What *Really* Causes Them?

". . . cursed is the ground for thy sake; in sorrow shalt thou eat of it all the days of thy life; thorns also and thistles shall it bring forth to thee; and thou shalt eat the herb of the field." Genesis 3:17–18

"The fact that bacteria are the cause of many infectious diseases is now so commonly accepted that it is difficult to visualize the scientific thought of the times when this relationship was unknown."[1]

"It is as causes of plant diseases that fungi have their greatest economic importance."[2]

"There are, however, many plants which usually grow where they are not wanted, have no economic value, and usually interfere with the production of cultivated crops or with the welfare of livestock. These plants are nearly always considered to be weeds."[3]

"Most types of plants, including all sorts of growing crops, are attacked and injured by insects."[4]

"Because modern insecticides are both effective and reliable, we in America are resorting to them more and more for the solution of our many insect problems."[5]

These quotations from standard college textbooks make it sound as if weeds and pests are inevitable and that the *cause* of diseases is germs, pathogens. Pesticides will solve our problems.

[1]C. L. Wilson & W. E. Loomis, *Botany*, 4th ed., 1967, p. 415.
[2]*Ibid.*, p. 429.
[3]D. Isley, *Weed Identification and Control in the North Central States*, 1960, p. 1.
[4]D. J. Borror, D. M. DeLong & C. A. Triplehorn, *Introduction to the Study of Insects*, 5th ed., 1981, p. 13.
[5]R. E. Pfadt, *Fundamentals of Applied Entomology*, 1962, p. 159.

Questions. Are our crops and animals *always* attacked by diseases and pests? Do weeds *always* infest our fields? Are toxic pesticides the *only* way of fighting these enemies? What *really* causes these problems, anyway?

Answers. We are more familiar with how our bodies work. We know that we have a built-in *immune system* which protects us from diseases and germs—well, most of the time. Actually, we have several different protective mechanisms, including our skin and mucus membranes as protective barriers, and internal antibodies and white blood cells as the defensive army. You may have heard of the recently discovered substance *interferon* which can defend against virus attack.

Stress. But you also realize that your natural defenses can break down. If you don't get enough sleep or exercise, or don't eat a nutritious, balanced diet, you are likely to come down with a cold or the flu. These *additional stresses* overtax the body's functions and ability to resist disease germs, and the germs, which were there all the time, are *allowed* to get a foothold. They only attack and cause the disease symptoms when the body is stressed and not in top-notch health.

The key. That is the key to understanding and preventing pest and disease attack. Not all animals and not all plants are attacked by insects or diseases. Sometimes one field will be devastated while the field across the fence remains perfectly healthy. Sometimes several plants in a row are healthy while the rest are attacked.

Both plants and animals have natural mechanisms of resistance to diseases and pests. If they have no serious genetic defects, *it is only when crops and animals are under severe stress that the diseases and pests are able to successfully attack*. They weren't in good health anyway, even before the germs or pests invaded. They were stressed to such a degree that their internal functions were beginning to be abnormal or inadequate. *They were already sick* (subclinically sick, to use the technical term). They would not have produced good quality crops or meat or milk. *They were inferior.*

Why diseases and pests? If you look at the over-all design of nature, the diseases and "pests" really have a useful function. *They are here to eliminate the sick, the weak, the unfit.* This is more easily understood by looking at a "wild" ecosystem, where many species of plants and animals live together and keep each other in check. Predators such as wolves can only catch and eat those animals that they are able to catch. Generally it is the sick or lame deer or moose that becomes a meal.

The reason pest and disease problems are so prevalent in agriculture today is that man inadvertently does things that put additional stresses on his crops and animals. He uses fertilizers and chemicals that upset soil ecology and prevent plants from getting the nutrients they need. He gives animals nutrient-deficient feed or confines them in poorly-ventilated buildings. He plants hybrid varieties that do not build high quality protein. Sometimes maybe it isn't inadvertent; sometimes it is done deliberately in the name of efficiency and profits. In any case it is short-sighted and unwise.

Pests and diseases actually provide a valuable service to us. Besides eliminating the unfit, they show us something is wrong, if we have the wisdom to heed their warning.

Many experiments have shown that insect pests and diseases attack crops that were given out-of-balance fertilization, such as too much nitrogen, or not enough phosphorus, potassium, or minor elements.[6]

Weeds too. The same basic principle applies to weed problems. In general, weeds grow best on poor, out-of-balance soil. That is their function in nature, to cover bare spots and waste places and prevent erosion. When they die they help build humus. Their roots bring up minerals (especially trace elements) from the subsoil and can break up a hardpan and open up tight soil. Some of them even make good forage for livestock.[7] Weeds are hardy "opportunists," always ready to colonize poor soil.

Some weeds grow best on clay (quackgrass), others on sand (sandbur), some on acid soil (sorrel, dock), others on saline soil (foxtail barley). Some take over on tight, anaerobic, or waterlogged soil (foxtail, bindweed, velvetleaf [buttonweed], Johnsongrass, jimsonweed). Thus weeds can be indicators of problem soils. Sheep sorrel is a classic indicator of low-lime soil. Dandelion and plantain grow well in alkaline soil, while common sunflower and giant ragweed (horseweed) do not.[8] A few weeds such as lamb's quarters and redroot (rough) pigweed grow best on good soil[9], and purslane prefers fairly good soil.[10]

Prevention or cure? When you get sick, you take whatever medicine is necessary to make you feel better. Actually the medicine does not cure you—your natural body processes do—but the medicine, plus rest and proper diet, assist the body to cure itself.

When your crops or animals get sick or are being ravaged by pests, you may have to resort to emergency measures and spray pesticides or have the vet give a drug. *Rescue chemistry*, it's sometimes called. You

[6]J. G. Rodriguez, p. 149–167 in *Biological and Chemical Control of Plant and Animal Pests*, 1960, L. P. Reitz, ed.

[7]J. A. Cocannouer, *Weeds, Guardians of the Soil*, 1950, p. vii–viii.

[8]A. S. Crafts & W. W. Robbins, *Weed Control*, 3rd ed., 1962, p. 89.

[9]C. Walters, Jr. & C. J. Fenzau, *An Acres U. S. A. Primer*, 1979, p. 322–324.

[10]E. E. Pfeiffer, *Weeds and What They Tell*, 1981, p. 50.

Indicator Weeds

Here is what some of the common U. S. weeds tell about soil conditions (scientific names are in parentheses):

- **bindweeds** (*Convolvulus*)—crusted, tight soil, low humus
- **broomsedge** (*Andropogon virginicus*)—depleted, oxidized soil, low in calcium and possibly magnesium; poor soil structure; possible overuse of salt fertilizers
- **foxtail barley** (*Hordeum jubatum*)—wet soil, possibly high salts and low calcium, compaction, possibly acid, unavailable potassium and trace elements
- **common burdock** (*Arctium minus*)—high iron, acid, low calcium; also grows on high gypsum soil or from excess use of dolomite lime or ammonium sulfate plus lime
- **cheat, chess** (*Bromus secalinus*)—wet, compacted, puddled (fine particles, no crumb structure)
- **chickweed** (*Stellaria media*)—high organic matter at surface, low mineral content
- **chicory** (*Cichorium intybus*)—fairly good soil, clay or heavy soil
- **cocklebur** (*Xanthium pennsylvanicum*)—fairly good soil with high available phosphorus, but may have low available zinc
- **crabgrass** (*Digitaria sanguinalis*)—tight, crusted soil, low calcium, inadequate decay of organic matter
- **dandelion** (*Taraxacum officinale*)—low calcium, organic matter not decomposing
- **dock** (*Rumex*)—wet, acid soils
- **fall panicum** (*Panicum dichotomiflorum*)—anaerobic, compacted soil
- **foxtail, giant foxtail** (*Setaria*)—tight, wet soil, possibly high magnesium; seed germinates in anaerobic conditions (high carbon dioxide)
- **horsenettle** (*Solanum dulcamara*)—crusted soil, low humus
- **jimsonweed** (*Datura stramonium*)—improper decomposition of organic matter (fermentation)

continued

have to get a crop that year or go bankrupt. It shouldn't be that way, but those are the realities of modern agriculture.

But is that the best approach? Isn't it much better to *prevent* the disease or problem from occurring in the first place? Of course it is. It saves yields, trouble, and money. And that's just the beginning. Now some people's idea of prevention is to spray the poisons ahead of time, just *in case* the bugs come, or because the weed seeds are going to sprout anyway. But this is only a band-aid approach, attacking the *symptoms* of a problem and not the *real causes*.

continued

- **Johnsongrass** (*Sorghum halepense*)—depleted soil, low organic matter, low calcium, possibly high iron
- **lamb's quarters** (*Chenopodium album*)—rich, fertile soil; good decay of organic matter, high humus
- **common milkweed** (*Asclepias syriaca*)—good soil, generally grows in fallow areas
- **mustards** (wild mustard, yellow rocket, wild radish, peppergrass, etc.) (*Brassica, Raphanus, Lepidium*)—crust, hardpan, poor soil structure, poor drainage
- **nettles, stinging nettle** (*Urtica*)—anaerobic, toxic soil, wrong decomposition of organic matter (fermentation)
- **pigweeds** (*Amaranthus*), **redroot (rough) pigweed** (*Amaranthus retroflexus*)—good soil
- **purslane** (*Portulaca oleracea*)—fairly good soil
- **quackgrass** (*Agropyron repens*)—wet, anaerobic soil, high aluminum (toxic); in West, low calcium and high magnesium and sodium
- **red sorrel, sheep sorrel** (*Rumex acetosella*)—acid soil, low calcium, low decomposition of organic matter
- **Russian thistle** (*Salsola kali* var. *tenuifolia*)—salty soil (high sodium and potassium), low calcium and iron, low organic matter
- **smartweeds** (*Polygonum*)—wet, poorly drained soil
- **thistles** (*Cirsium*) & **sowthistle** (*Sonchus oleraceus*)—fairly good soil
- **tumbleweed** (*Amaranthus albus*) (Russian thistle is also called a tumbleweed; see above)—dry soil, low humus
- **velvetleaf (buttonweed)** (*Albutilon theophrasti*)—anaerobic soil, wrong decay of organic matter (fermentation)

Compiled from: C. Walters, Jr. & C. J. Fenzau, 1979, *An Acres U. S. A. Primer*, p. 302–349; E. E. Pfeiffer, 1981, *Weeds and What They Tell*, p. 12–16, 33–93; S. B. Hill & J. Ramsay, 1977, *Weeds as Indicators of Soil Conditions*.

Diseases, weeds, and pests can really be prevented and the problems solved by eliminating what causes them. Diseases and pests attack when plants or animals are under too much stress and their natural defenses are down. Weeds are a problem on anaerobic and out-of-balance soil. Before we go into detail on solving these problems, let's look at the results of the other approach—the use of medicines and pesticides.

What chemicals do. What do toxic chemicals actually do? How do medicines work? Do they have any other effects besides killing germs, bugs, or weeds?

The basic purpose for using pesticides and medicines is to *kill*, whether it is germs, weeds, or bugs. Therefore these chemicals are usually at least toxic or poisonous to the "target organisms"—those species we want to kill.* A few are pretty specific and only kill one or a few species, but the great majority can kill other non-target organisms. So in the case of insects, insecticides not only eliminate, say the greenbug aphid, but also the predators and parasites that help keep it under control—if they are given a chance. Green lacewings, ladybugs, and tiny parasitic wasps are efficient natural enemies of aphids. Or how often does herbicide drift or carryover from last year injure a crop? *These are undesirable side effects.*

Many medicines (drugs) are pretty specific for a certain type of pathogen, but like pesticides, *they have side effects.* They disturb or inhibit some body function. They put additional stress on the body, which can weaken an animal so much that it comes down with another disease.

Insecticides kill insects in a variety of ways. Some have to be eaten to kill (stomach poisons), some act when breathed into the respiratory system (fumigants), and some kill when any part of the insect's body touches them (contact poisons). The actual death of the pest can be caused by cell destruction of certain internal tissues. Others attack the nervous system. DDT causes nerve cells to "fire" repeatedly. Organophosphate insecticides such as parathion were developed from nerve gas research during World War II. They suppress vital enzymes, cholinesterases, which regulate nerve impulses.[11]

Herbicides kill plants by such mechanisms as interfering with various cellular metabolic functions (amino acid production, energy transfer, cell membrane functions), destroying photosynthesis capability, decreasing stability of the cell's nucleus, and interfering with cell division. The well-known 2, 4 - D acts as a synthetic plant hormone, resulting in abnormally slow tip growth and sometimes tumors.

Side effects. Many herbicides and fungicides are absorbed into the plant's tissues and carried (translocated) throughout the plant. They can end up deep in the soil by going through the roots. They can enter crop plants and end up in animal feed and the fruits and vegetables we eat. In vineyards and orchards, where fungicides containing heavy metals are used frequently, copper, zinc, and manganese levels have been increasing in apples and grapes. Apples from sprayed orchards in Europe contained 3.5 times the zinc, 2.4 times the copper, and 5 times the manganese as unsprayed apples.[12] Most pesticides are toxic to at least a few

*Actually, there are a few pesticides that kill "bugs" but are not poisonous, such as diatomaceous earth, which abrades insects' outer cuticles and they die from drying out; or attractant odors that lure insects into traps; large raindrops will kill aphids just by mangling them.

[11]H. Wellford, *Sowing the Wind*, 1972, p. 222.

[12]D. D. Kerin, *Efficient Use of Fertilizers in Agriculture*, 1983, p. 352.

plant species, and soil fumigants are toxic to most plants. If crops are planted too soon after application, if soil is cold and wet, or if pesticide residues have built up over the years, crop failure can result.[13]

Pesticides can affect the crop plant's functions, leading to abnormal levels of nutrients, plus lower quality and flavor. In beans, 2, 4 - D caused lowered protein content, while in sugar beets it caused such a high nitrate content that they would be toxic if eaten by cattle.[14] The soil insecticide benzene hexachloride (BHC) causes off-flavors in many fruits and vegetables, as does lindane. Some fungicides and insecticides used on cherries cause poor flavor and quality (color, size, weight, acidity).[15]

Almost nothing is known about the effects of combinations of two or more pesticides. In some cases a combination is much more toxic than either chemical is alone (called a synergistic effect).[16] The nerve poison aldicarb (TEMIK) is an extremely toxic insecticide (a few drops on your skin can be fatal). In the environment, it rapidly oxidizes to form aldicarb sulfoxide, which is 47 times more toxic! Dr. Donald Kaufman, a USDA microbiologist, found that in the soil, certain pesticides block the breakdown of other pesticides. Aldicarb is similar to other chemicals that have been found to act that way.[17]

Many herbicides and insecticides are volatile, evaporating quickly into the air. Others evaporate more slowly, but on a hot day the rate speeds up greatly. This may be good in that it reduces soil contamination by poisons, but does it create a toxic air pollution and a deadly "poison rain," similar to acid rain, miles away? Evidence indicates that it does. Pesticide evaporation, plus airborne spray particles and dust can literally be carried around the world by upper air currents. Atrazine, 2, 4 - D, DDT, DDE, BHC, and dieldrin have been found in rain in such places as England and Hawaii. Extremely high levels of aldrin, chlordane, toxaphene, and DDT were measured in the air over agricultural areas of Florida and Mississippi in the 1960's. When the volatile thiocarbamate herbicide EPTC is applied to soil in water solution, 70% of it evaporates the first day if the soil is moist (45% if it is dry).[18]

Soil problems. When pesticides get into the soil (some are applied directly to or into the soil, some drip off plants or are washed into the soil

[13]J. P. Martin, p. 754–755 in *Organic Chemicals in the Soil Environment*, Vol. 2, 1972, C. A. I. Goring & J. W. Hamaker, eds.

[14]C. J. Willard, *Proc. North Central Weed Control Conference*, 1950, vol. 7, p. 110–112.

[15]C. H. Mahoney, *Residue Reviews*, 1962, vol. 1, p. 11–23.

[16]M. W. Miller & G. G. Berg, *Chemical Fallout*, 1969, p. 440.

[17]*The New Farm*, September/October 1983, p. 30–31.

[18]A. W. A. Brown, *Ecology of Pesticides*, 1978, p. 299–302, 332.

by rain), many undesirable and unexpected results can occur. As we have mentioned in Chapter 3, some soil organisms are killed immediately, while others are fairly tolerant.

The insecticide heptachlor is toxic to nitrifying bacteria at 50 ppm and legume nitrogen-fixing bacteria at 100 ppm.[19] Carbofuran at 4 lbs./acre killed 95% of the earthworms in one study, and disulfoton used for aphids on cotton at 2 lbs./acre in the furrows killed 95% of certain soil arthropods (collembola and mites). Diazinon at 8 lbs./acre killed 64% of beneficial predatory ground beetles, while parathion at 8 lbs./acre killed 95% of them.[20]

Fungicides and soil fumigants are the worst, killing large numbers of soil organisms.[21] It may take anywhere from weeks to up to a year before sterilized soil is recolonized by normal populations of organisms. Plant diseases can be worsened if a fungicide kills off beneficial microbes that normally hold the pathogenic species in check, although in some cases, beneficial disease-fighting species proliferate after soil treatment.[22] Pesticides can cause mutations of some soil organisms (*Aspergillus* fungi).[23]

Most pesticide studies are only concerned with short-term effects. In one long-term study, yearly applications of 3.56 lbs./acre of atrazine to kill weeds in an orchard caused significant soil changes after 14 years. Humus content of the weed-free soil had dropped to nearly half that of untreated soil, and pH was lower (4.7 in atrazine soil, 5.3 in untreated). Total numbers of soil bacteria and fungi were about the same, BUT the balance was greatly upset: some humus-forming bacteria and anaerobic species were permanently reduced, while other humus-formers and denitrifying bacteria were only temporarily reduced by atrazine. Other types (ammonifying and nitrogen-fixing) were increased. The normal nitrogen cycle in the soil (nitrification: change of ammonium to nitrate nitrogen) was seriously reduced.[24]

Herbicides, fungicides, and soil fumigants have been found to release plant nutrients in the soil by killing some microbes and stimulating others, from breakdown of the pesticide, and by direct chemical action of the pesticide on the soil.[25] If the plants can't use all of the released

[19]*Advances in Agronomy*, 1971, vol. 23, p. 212.

[20]A. W. A. Brown, *ibid.*, p. 68, 72, 84.

[21]J. P. Martin, *ibid.*, p. 737–741; J. C. Headley & J. N. Lewis, *The Pesticide Problem*, 1967, p. 78; J. P. Martin, *Residue Reviews*, 1963, vol. 4, p. 103–104.

[22]J. P. Martin, *ibid.*, 1963, p. 106.

[23]*Advances in Agronomy*, *ibid.*, p. 213–215.

[24]J. P. Voets, P. Meerschman & W. Verstraete, *Soil Biology and Biochemistry*, 1974, p. 149–152.

[25]J. C. Headley & J. N. Lewis, *ibid.*; J. P. Martin, *ibid.*, 1972, p. 747–748; J. P. Martin, *ibid.*, 1963, p. 109–110, 117.

nutrients, some can be leached away. Some trace elements (manganese, copper, zinc) can be made more soluble to such an extent that they become toxic.[26] High levels of ammonia can build up to toxic levels (to sensitive plants) because of the high numbers of dead soil microbes and inhibition of nitrifying bacteria.[27]

Although most pesticides apparently do not harm soil tilth directly, the killing of humus-producing organisms and the reduced amount of plant residues returned to the soil (from herbicides) can lessen humus levels and degrade soil structure in the long run. Also, the high levels of ammonia sometimes produced can destroy tilth more quickly.[28]

Following application of soil fumigants, fungicides, or other treatments that kill soil organisms, crop growth is often inhibited, and plants do not take up enough phosphorus, zinc, or copper. The problem may be caused by toxins produced by excess growth of certain microorganisms, while beneficial species that would have destroyed the toxins have been eliminated. Similarly, disease pathogens can get the upper hand because their natural enemies have been destroyed. Beneficial mycorrhizae fungi are killed by many fungicides.[29]

Although not sold as a pesticide, a recently-developed chemical (nitrapyrin) is called a nitrogen stabilizer and is applied to soil to inhibit the nitrification process by killing *Nitrosomonas* bacteria. Thus ammonium is not converted to nitrite (and then to nitrate), and nitrogen stays in the soil for a longer time. Research has found that nitrapyrin and its breakdown product, 6-CPA, have relatively low toxicity for most other forms of life in soil, water, and on land (although few species have been tested). At high concentrations (100–1000 ppm) soil fungi are inhibited, however. If soil and weather conditions are not right, there may be no crop yield increases, according to an article in *Down to Earth* (1976, vol. 32, no. 1, p. 22–26).

In the ecosystem. The problems that result when pesticides become distributed in the environment are well known. Very small amounts end up in streams and lakes, get into algae, and then into crustaceans that eat the algae. Next into fish and then into birds and mammals. Then into humans, to add to the pesticide residues we get directly from the vegetables, fruits, and meat we eat. At each step up the food chain, the amount of pesticide is concentrated, so while lake water may only contain 0.000001 part per million DDT, crustaceans may have 0.001 ppm, fish

[26]J. P. Martin, *ibid.*, 1963, p. 110, 112.
[27]*Ibid.*, p. 112–113.
[28]*Ibid.*, p. 111; J. P. Martin, *ibid.*, 1972, p. 750; A. W. A. Brown, *ibid.* p. 64.
[29]J. P. Martin, *ibid.*, 1972, p. 763–768.

and birds 2.0 ppm, and humans 6.0 ppm.[30] Nearly every person tested by the Department of Epidemiology, Miami University, Miami, Florida, had DDT, dieldrin, lindane, and other insecticides in their body fat. Also, 100% of the people had pentachlorophenol in their urine, a by-product of pesticides which causes fetal tumors in rodents.[31]

Dr. Americo Mosca of Turino, Italy, has found that toxic agricultural chemicals cause damage to tissues by ionization, leading to mutations, birth defects, and cancer, just as radiation from atomic fallout.[32]

Resistance. Probably the most ironic side effect of the use of pesticides is that the very pests they were supposed to eradicate have bounced back even stronger than before. They have developed resistant varieties, which are immune to the old pesticides, and the chemical companies are kept scrambling to come out with new, stronger poisons. We now have "superflies" and potato beetles that have to be sprayed weekly rather than once or twice a year. The number of insects resistant to one or more insecticides has shot up from about 25 in the mid 1950's to over 400 in 1980. There are over 30 weed species resistant to such herbicides as atrazine, Bladex, and Princep. Since 1970, at least 90 species of fungicide-resistant plant diseases have been found. The problem is occurring world-wide.[33]

Herbicides can even worsen insect problems. In one test of the effect of 2, 4 - D use on corn, 16% of untreated plants had corn borers, compared to 28% and 24% for normal and 1/4 normal doses of 2, 4 - D. Numbers of corn leaf aphids per plant were 618 (untreated), 1679 (normal dose), and 1388 (1/4 dose).[34]

Some fungus diseases are increased by application of herbicides; early blight of tomato and mosaic virus of tobacco, cucumber, and cotton are increased by 2, 4 - D, while damping-off of cotton increases when trifluralin is used.[35]

Conclusions. One over-all observation that can be made about the use of synthetic chemicals as a "quick fix" for agricultural problems is that *they interfere with nature*. They do not work *with* natural laws and processes—they try to overpower nature. And as we are finding out, it just doesn't work!

We are going to have to swallow our pride and admit we aren't as smart as we thought we were. We need to study nature more closely,

[30]C. A. Edwards, *Persistent Pesticides in the Environment*, 1970, p. v.
[31]*Acres U. S. A.*, November 1981, p. 7.
[32]*Acres U. S. A.*, February 1982, p. 33.
[33]*Science*, 1984, vol. 226, p. 1293; *The New Farm*, May/June 1984, p. 19, 21.
[34]I. N. Oka & D. Pimentel, *Science*, 1976, vol. 193, p. 239–240.
[35]A. W. A. Brown, *ibid.*, p. 325.

discover (or rediscover) its laws and processes, and put them to work for us. That's how things were intended to work.

Natural protection. Let's take a look at how nature protects plants from diseases and pests. First, you should realize that disease pathogens and pests, in general, are always around, at least in the warm months. Yet not all plants come down with diseases or are attacked by pests. So diseases and pests are not inevitable, and the germs and bugs are not the real causes of the problems.

They are just there, ready and waiting, to eliminate any inferior, unfit, unhealthy plant which is weakening under stresses. Food from that plant would not be good for man nor beast, so the pests and pathogens are trying to do us a service by eating them before we can. Most of us, however, do not understand this and zap them with poison and eat the inferior food anyway. And then we wonder why we get sick or die of heart attack at age 45.

Pathogens and pests do not successfully attack healthy plants because, just as animals and humans do, they have various methods of repelling, resisting, and killing the attackers. New ones are being discovered every year.

Natural resistance. Methods and factors that contribute to natural resistance to diseases and pests are as follows:

1. *Barriers.* Plants are covered on the outside by a layer of waxy cuticle. It not only protects from loss of water from the cells, but if it is thick, it can protect from some pathogens.[36] Some plants have a dense mat of hairs or spines on their leaves or stems, which can be an effective barrier to insects. In some species or varieties, the hairs have tiny hooks at the tip or release a sticky secretion. These are perfect traps for small insects, such as aphids, leafhoppers, flea beetles, and young corn borer larvae. Research now is being done to develop special varieties of hairy potatoes, alfalfa, and probably other crops.[37]

2. *Repellent chemicals.* Some plants produce strong-smelling odors or toxic chemicals which keep pests from even starting to attack. Generally, if they smell strong or unpleasant to us, they will repell insects. Examples include turpentine and resins found in conifer trees, mint oils, citrus oils, and garlic and onion oils. The use of marigolds, garlic, chrysanthemums, anise, nasturtiums, and other plants to protect vegetables from pests is well-known to many gardeners.

A certain species of wild potato has been discovered to produce a chemical that is similar to the odor (called an alarm pheromone) that

[36]F. B. Salisbury & C. Ross, *Plant Physiology*, 1969, p. 382.
[37]*The New Farm, ibid.*, p. 22–23.

aphids use to alert other aphids that there is danger nearby. Thus, aphids don't even come near the plants.[38]

3. *Phytoalexins.* If bacteria invade a leaf and kill a few cells, or if a fungus spore germinates, some plants secrete chemicals called phytoalexins, which kill the pathogens or inhibit germination of more spores. Corn produces two phytoalexins that give resistance to the northern corn leaf blight fungus.[39] Phytoalexins have been found in dozens of species of plants and give resistance to fungi, bacteria, viruses, and nematodes.[40]

4. *Amino acids and proteins.* Certain proteins and amino acids that are not part of proteins are found in resistant varieties of plants. [41] Treating susceptible apple trees with the amino acid phenylalanine markedly increased their resistance to apple scab disease. [42] Both synthetic and natural amino acids can control certain diseases. For example, cysteine, cystine, leucine, and methionine inhibit cucumber scab; and serine, threonine, isoleucine, cysteine, histidine, methionine, glutamic acid, arginine, and tryptophane were effective against wheat rust in lab tests. More effective disease control results when there is a proper balance of amino acids, vitamins, and trace elements. Excessive fertilization with inorganic nitrogen salt fertilizers and urea increases susceptibility of many plants to fungal diseases, and high potassium increases late blight in potatoes because of unbalanced amino acids in the plants.[43]

5. *Other plant chemicals.* Plant cells produce an amazing variety of chemicals as by-products of metabolism, sometimes called secondary metabolites. The functions of many of them are unknown, but some can make the plant resistant to pests or diseases. Some of them are just plain lousy-tasting to insects. Others cause insects to stop eating or reduce their fertility. Some inhibit the insects' digestive enzymes so that they are not able to digest their food. Some inhibit the growth of fungi or bacteria.

These metabolic chemicals include tannins, protocatechuic and chlorogenic acids, coumarin derivatives, proteinase inhibitors, and polyacetylenes. Some are only produced when the plant is injured or attacked by pests or disease pathogens.[44]

[38]R. W. Gibson & J. A. Pickett, *Nature*, 1983, vol. 302, p. 608–609.

[39]S. M. Lim, A. L. Hooker & J. D. Paxton, *Phytopathology*, 1970, vol. 60, p. 1071–1075.

[40]J. A. Bailey & J. M. Mansfield, *Phytoalexins*, 1982.

[41]G. A. Strobel & E. L. Sharp, *Phytopathology*, 1965, vol. 55, p. 413–414; J. A. Goss, *Physiology of Plants and Their Cells*, 1973, p. 201–204; *Science News*, Nov. 2, 1974, p. 286.

[42]J. Kuć, E. Barnes, A. Daftsos & E. B. Williams, *Phytopathology*, 1959, vol. 49, p. 313–315.

[43]O. M. van Andel, *Annual Review of Phytopathology*, 1966, vol. 4, p. 349–368.

[44]*Science*, 1982, vol. 216, p. 722–723; 1984, vol. 225, p. 53-54; F. B. Salisbury & C. Ross, *ibid.*, p. 393–394.

A newly developed variety of sweet potato is resistant to rootknot nematodes, fusarium wilt, and ten species of insects, including wireworms, grubs, and flea beetles. A wild tomato from Ecuador has a substance in its hairs that kills nearly 100% of whiteflies, flea beetles, tomato and tobacco hornworms, and Colorado potato beetles. [45] Chemicals found in corn, including one called DIMBOA, inhibit or repell young European corn borers and spores of northern corn leaf blight and gibberella stalk and ear rot.

6. *Genetic varieties.* It is well known that certain varieties of plants are more resistant to diseases and pests than others and that this resistance is controlled by one or more genes. Thus, modern agriculture has spent very much research effort to discover and develop resistant genetic varieties. The exact mechanism of resistance may be one of those previously discussed, such as repellent odors, barriers, phytoalexins, or other plant chemicals.

By "building in" the resistance to a variety of hybrid seed, the natural mechanism is automatically made available to the farmer. Varieties of wheat resistant to wheat rust and Hessian flies, soybeans resistant to phytophthora root rot, alfalfa resistant to verticillium wilt and aphids, and corn resistant to European corn borer are examples.

7. *Overall health of plant.* A stressed plant's functions are out of kilter. Some of the hundreds of internal cell and tissue functions will be abnormal or inhibited. Under anaerobic soil conditions such as result from flooding, roots do not absorb nutrients, imbalances in hormones occur, cellular metabolism shifts to fermentation, toxic levels of alcohol can build up in tissues, etc. Disease pathogens readily attack such stressed plants.[46]

Stressed plants may not be able to produce their normal resistance-chemicals, such as phytoalexins or amino acids. Stressed plants even look and smell different to insects, which is how insects locate sick plants to attack. Fascinating research that proves this has been done recently by USDA researcher Philip S. Callahan. All objects, including plants, radiate ("broadcast") invisible infrared wavelengths depending on their temperature and chemical make-up. Odor molecules broadcast at different wavelengths depending on their type, concentration, temperature, and vibrations. Some of the tiny hairs on the antennae of insects are sensitive to infrared wavelengths. Stressed plants are abnormal and have different odors and higher temperatures than healthy ones.[47] Thus sick plants "look" different to insects from healthy ones, and the bugs can

[45] *The New Farm, ibid.*

[46] R. D. Durbin, p. 102–107, 109, 111 in *Water Deficits and Plant Growth*, Vol. V, 1978, T. T. Kozlowski, ed.

[47] *Acres U. S. A.*, September 1982, p. 7; *Successful Farming*, December 1981, p. 26–P.

easily home in on them. In fact, that is one reason mosquitos can easily locate you at night: your skin "glows" at infrared wavelengths. For more details, see Dr. Callahan's books, *Tuning in to Nature* (1975) and *Ancient Mysteries, Modern Visions* (1984).

8. *Soil life.* As we have mentioned in Chapter 2, several types of soil organisms directly or indirectly protect plants from diseases and pests. A whole host of beneficial organisms act together in healthy soil to produce a proper balance of plant nutrients and growth factors, thus producing a vigorous plant that will be more resistant. More specifically, certain fungi (mycorrhizae, *Penicillium*, etc.), actinomycetes, and bacteria (especially those near roots, in the rhizosphere) directly attack disease pathogens by releasing antibiotics or by mechanisms that starve them to death.[48] Certain species of fungi actually trap and eat nematodes.[49]

One of the best ways to increase beneficial soil organisms is to add organic matter (green manures, animal manures, compost).[50] In one study, a California grape grower who had nematode problems, even after using the toxic nematicide Nemagon, applied three tons of compost per acre. The nematode problem ceased, soil structure improved, and yields started up.[51]

Tests at the University of Illinois found that inoculation of soybean seeds with certain strains of rhizosphere bacteria protected plants from phytophthora root rot just as effectively as plants treated with a fungicide, and out-yielded untreated plants by 129%.[52] Similar inoculation of potato seed pieces increased yields up to 33%, while radishes increased in root weight 60 to 144% over untreated plants.[53]

Plant pathologist R. J. Cook believes that one way humus and soil microorganisms help control diseases is by small amounts (a few parts per million) of ethylene gas, produced by some anaerobic soil bacteria. Ethylene, which is also produced in plants, where it functions as a plant hormone, inhibits the germination of disease fungus spores (such as *Fusarium* and *Sclerotium*). The ethylene is produced especially when organic matter is added to the soil, providing food for the bacteria, while excessive tillage and high nitrate levels decrease ethylene production.[54]

9. *Soil nutrients.* The proper amounts and *balance* of nutrients must be present in the soil in order for plants to be healthy. If one element is

[48]K. F. Baker & R. J. Cook, *Biological Control of Plant Pathogens*, 1974, p. 170–200; M. N. Schroth & J. G. Hancock, *Science*, 1982, vol. 216, p. 1376–1381.

[49]G. L. Barron, *The Nematode-destroying Fungi*, 1977.

[50]K. F. Baker & R. J. Cook, *ibid.*, p. 81–85, 333–336.

[51]*Acres U. S. A.*, September 1984, p. 1, 6.

[52]*Farm Journal*, August 1982, p. D–8.

[53]M. N. Schroth & J. G. Hancock, *ibid.*, p. 1378.

[54]*Compost Science*, 1976, vol. 17, no. 2, p. 23–25.

out of balance (too high or too low), others can be affected. Deficiencies can result. High or low pH can also restrict nutrient availability and cause deficiencies. Some of the interrelations of mineral nutrients and pH are shown on the accompanying charts.

Nutrient Interactions

If the element (or soil condition) at the left is too high (H) or too low (L), a deficiency of the element at the top can result.

(information compiled from many books and journals).

	N	P	K	Ca	Mg	S	Fe	Al	Cu	B	Zn	Mo	Mn	Na	Cl	Co
N (nitrogen)			H							H	H					
P (phosphorus)	L		H	H			H		very H	H	L	H				
K (potassium)				H	H					H	very H					
Ca (calcium)		H			H		H			H						
Mg (magnesium)	H	H	H	H												
S (sulfur)	L											H	L			
Fe (iron)		H + Al + lowpH		H + Al,Mn + lowpH	H + Al,Mn + lowpH				H + Mn							
Al (aluminum)		H + Fe + lowpH		H + Fe,Mn + lowpH					H							
Cu (copper)							H					H				
B (boron)			H	L	H											
Zn (zinc)							H									
Mo (molybdenum)	L						H		H							
Mn (manganese)	H		H + Fe,Al + low pH	H + Fe,Mn + low pH			H + low pH		H + Fe							
Na (sodium)			H	H												
Cl (chlorine)			H													
Co (cobalt)																
Ni (nickel)							H									
pH		L + H Fe,Al		L + Al Fe,Mn	L + Al, Fe,Mn		H or L + Mn		H	H or L	H	L	H + org. m.			H
organic matter									H	L	H		H + H pH			

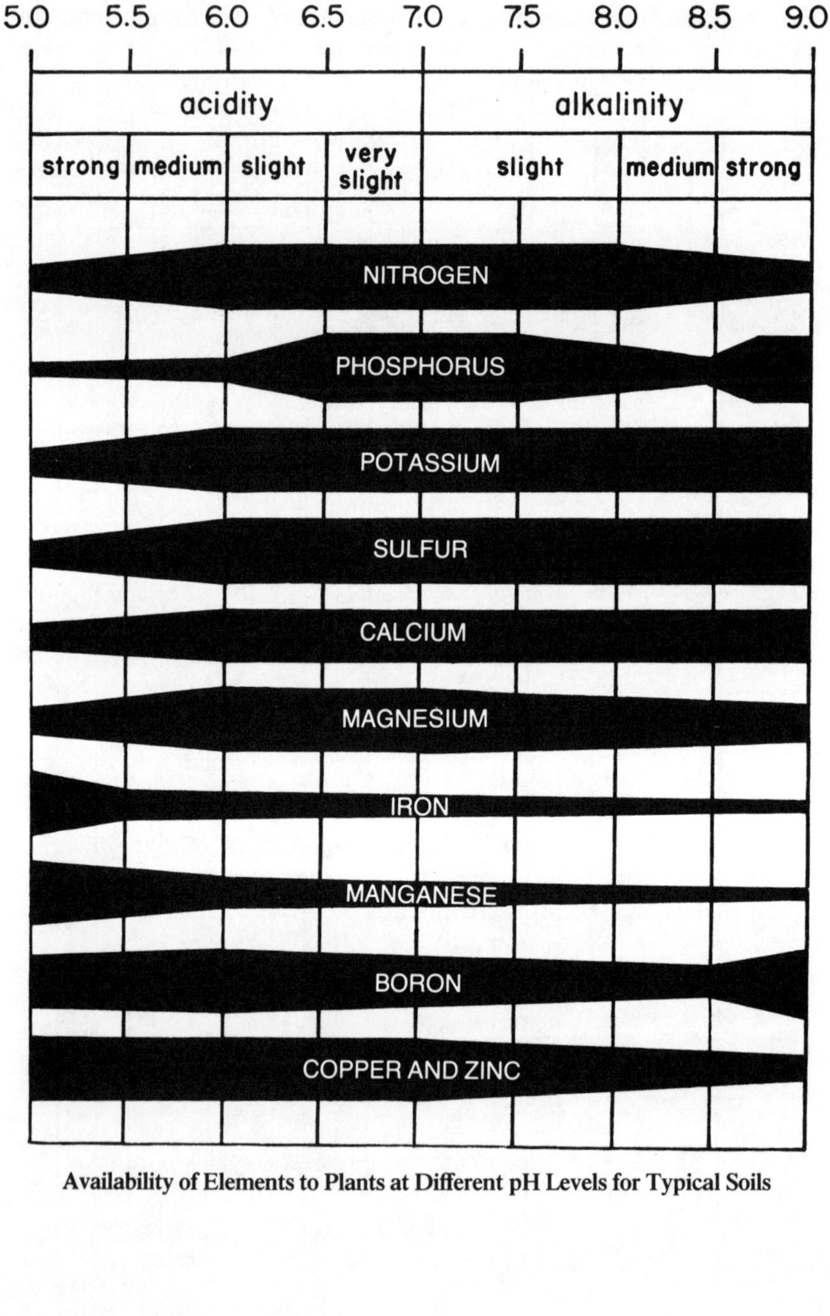

Availability of Elements to Plants at Different pH Levels for Typical Soils

For example, calcium and magnesium are "antagonists," "competing" for root absorption, so to speak. If there is too much magnesium relative to calcium, the plant will be deficient in calcium, and unhealthy cell walls, poor root development, slow growth, and poor crop quality can result. Increased calcium levels in potatoes reduce Erwinia soft rot up to 50%.[55] High potassium can cause calcium, magnesium, boron, iron, and zinc shortages; and when potassium chloride (muriate of potash) is used, production of amino acids, proteins, chlorophyll, and vitamins C, B_1, and A (carotene) will be reduced; crop quality will go down.[56] Excess nitrogen fertilization is notorious for increasing disease and pest vulnerability in plants. *Many* experiments have proved this. For example, high nitrogen, as well as surface mulch (as would occur in reduced tillage systems) increased corn borer infestation, possibly because a corn borer inhibitor chemical produced in the rind of the corn stalk is reduced.[57]

Excess nitrogen produces a fast-growing, but weak and watery plant. In one experiment, high nitrogen fertilization decreased the numbers of soil fungi that fight diseases, while a medium level of phosphorus increased them.[58] In an experiment on red clover and Fusarium diseases, "vigorously growing plants had the least amount of disease." [59] A moderate amount of nitrogen, as well as the proper amounts of phosphorus, calcium, potassium, and all the other elements, gives increased resistance. Balance is the important thing.[60] "Plant disease depends on a preceding, exactly defined, specific, mineral deficiency. . . plant diseases probably increase parallel with the soil deterioration and have their origin in the deficient or unbalanced plant nutrition."[61]

10. *Other soil conditions.* We covered the vital importance of well-aerated soil in Chapter 3. Anaerobic (non-aerated) soil and high toxin levels are certain to put stresses on plants, kill or inhibit beneficial soil organisms, and lead to disease and pest attack. Proper soil moisture level, pH, and temperature are also important: too high or too low creates stresses.

Grasshoppers are a problem where dry soil conditions exist. Corn

[55]*American Vegetable Grower*, September 1982, p. 62.

[56]D. M. Huber & R. D. Watson, *Annual Review of Phytopathology*, 1974, vol. 12, p. 151; D. Davidescu, p. 1095 in *Fertilizers, Crop Quality and Economy*, 1974, V. H. Fernandez, ed.; G. N. Null & S. Null, *How to Get Rid of the Poisons in Your Body*, 1977, p. 153.

[57]G. P. Offerman, A. J. Pappelis & J. P. Vavra, *Phytopathology*, 1967, vol. 57, p. 1031–1033.

[58]D. D. Kaufman & L. E. Williams, *Phytopathology*, 1965, vol. 55, p. 570–573.

[59]K. T. Leath & R. H. Ratcliffe, *Forage Fertilization*, 1974, p. 488.

[60]*Ibid.*, p. 482.

[61]A. Primavesi, p. 152 in *Fertilizers, Crop Quality and Economy*, 1974, V. H. Fernandez, ed.

borers attack where soil has poor structure, low humus, and low pH and calcium (or normal pH and high magnesium). Cutworms occur in acidic, low calcium soil. Wireworms like tight, wet soil (often because of high magnesium). Japanese beetles tend to occur where organic matter is not decaying aerobically.[62] Even slight soil compaction and restriction of drainage leads to severe infection of Fusarium root rot and other diseases in peas in eastern Oregon and Washington. A beneficial soil fungus that fights the disease pathogens is being studied by USDA researchers.[63] During the summer of 1984, the Wisconsin Cooperative Pest Survey Bulletin of June 29, July 27, and August 17, stated that armyworms and corn stalk borers were worse in no-till or minimum-till fields in southwestern Wisconsin.

How to foster natural resistance. What can you do in your fields to promote natural resistance to diseases and pests? The answers have been scattered throughout the preceding pages, but let's collect them together and summarize.

1. The overall thing to strive for is a vigorous, healthy, stress-free plant. In reality, there will always be some stresses, but try to reduce those you can control—primarily factors of the soil.

2. The soil should have good structure (tilth). This gives good aeration and drainage. An adequate humus level is essential for good structure. Tillage methods should be appropriate for your climate, soil type, and crops. If soil conditions are not ideal, rotation of crops can give good protection from certain pests, especially those that do not migrate far and those that mainly live in the soil, such as nematodes, wireworms, and gall gnats, as well as many fungal diseases.[64]

3. Soil fertility should be high and balanced. Frequent soil tests should be run, and proper corrective and/or maintenance fertilization should be done. An adequate humus level, regular recycling of organic matter, and a healthy population of soil organisms are the ideal ways of achieving this. High application of synthetic commercial fertilizers may increase pest or disease vulnerability.[65]

4. High quality seed and crop varieties should be grown, adapted to your climate.

5. Any materials applied to the soil should be free of (or not produce) toxins that may injure soil organisms or destroy humus. Pesticides

[62]C. Walters, Jr. & C. J. Fenzau, *ibid.*, p. 273–280.

[63]*Acres U. S. A.*, June 1983, p. 10–11.

[64]E. Zanini & F. Frilli, p. 715–717 in *Natural Products and the Protection of Plants*, 1977, G. B. Marini-Bettlo, ed.

[65]O. M. van Andel, *ibid.*, p. 356–357; J. G. Rodriguez, *ibid.*, p. 155–156, 158–159.

should only be used as a last resort. Even organic matter can produce toxins in the soil if not applied properly.

More details on how to do these things will be given in the next chapter.

Natural pest control. But if your soil isn't yet in healthy condition, or if bad weather puts your crops under stress, and if insects, mites, or nematodes do attack your plants—then what should you do? Drench them with poison?

Try to avoid that drastic option until you look at the situation more carefully. Of course, not all "bugs" on your plants are bad ones. There will be many "good guys"—natural enemies of the "bad guys"—such as green lacewings, ladybugs, damsel bugs, tachinid flies, and parasitic wasps, to name a few. Check out a "bug book" from the library and get familiar with your friends. Learn how to tell a helpful damsel bug from a harmful chinch bug.

Even if there are some pests chewing leaves or sucking sap, is the damage they are doing really significant? Most plants can lose up to 30–40% of their leaf area before they are harmed, and vigorous plants can sustain heavier attack and outgrow it. In a study of pea plants, the plants grown on nutrient-deficient soil suffered most from aphid attack, but the vigorous, well-nourished plants actually had more aphids on them. [66]

According to pest control experts, if pest damage is below a certain level, called the economic threshold, economic injury level, or critical level of damage, applying a pesticide or any other control will not pay. This varies depending on the value of the crop and the cost of the control. However, only short-term economics are considered; any long-term damage to the soil and natural enemies is ignored.

Let's say that pest damage is above the economic threshold or soon will be—what is the best control method? There is no simple answer; it depends on the situation—the crop, the pest, the weather, soil conditions, and whether there are natural enemies. Some careful field scouting and detective work is needed.

If the soil and weather conditions are favorable for vigorous plant growth, and if a survey reveals plentiful natural enemies (predators, parasites, or diseases), let nature take its course (while keeping a careful watch on the situation). The pests should be controlled by their natural enemies, within a couple weeks usually. If natural enemies are few or not doing their job, you can buy and release natural enemies to help nature along. This should be done about two weeks before anticipated serious pest damage.

[66] J. S. Baker & O. E. Tauber, *Jour. of Economic Entomology*, 1951, vol. 44, p. 1010–1012.

You can buy predators such as adult ladybugs (release 1/2 to 3 gallons per acre) or green lacewing eggs (10,000 to 50,000 per acre); these are good for small pests such as aphids, leafhoppers, scales, mealybugs, and mites. Three species of predatory mites (1,000 to 10,000 per acre) are available to kill plant-feeding mites on beans, roses, cucumbers, squash, peaches, strawberries, grapes, apples, and citrus. Praying mantis egg cases (10 to 100 per acre) hatch into young mantids that eat aphids, small caterpillars, etc., and then grow into large adult mantids that consume grasshoppers and other large pests. And you can put up birdhouses to encourage wrens, bluebirds, swallows, etc.

Parasitic wasps called *Trichogramma* can be bought in egg form; they parasitize eggs of corn earworms, cutworms, pink bollworms, leafworms, armyworms, codling moths, cabbage loopers, oriental fruit moths, peach tree borers, and other caterpillars. Parasitic wasps for alfalfa weevils, Mexican bean beetles, and certain other pests are being developed. Parasitic nematode worms can be sprayed or scattered in a cedar chip medium for control of corn rootworms, corn borers, cutworms, wireworms, black vine weevils, cabbage root maggots, and carpenter worms.[67]

A caterpillar bacterial disease called *Bacillus thuringiensis* (BT) can be sprayed, and a milky spore disease (another bacterial disease) is being used for white grubs and the Japanese beetle. A virus disease to control the codling moth is being developed in California.[68] When ordering natural enemies, state the pest(s) you want to control, since there may be different strains or species for different pests.

Collecting some of the insects that are damaging your crops, grinding them up in a blender with a quart of water, and (after straining) diluting with water and spraying on the plants has been found to effectively kill the pest species. Entomologist F. R. Lawson believes the method may work by spreading an insect disease virus, or possibly by a repellent odor. About a half-pound of insects will treat about 75 acres if about 4 oz. of the blender liquid is diluted with 1500 gallons of water.[69]

If you can't wait a couple weeks for natural enemies to do their job, there are other ways of killing pests that do not pollute the environment with highly toxic chemicals. For fruit trees or other trees, the use of a dormant oil spray in late winter or early spring or a summer oil spray can control (by suffocation) scales, and aphids. Ordinary sugar, worked into the topsoil will not only kill nematodes directly by drying them out, but also stimulates the growth of beneficial soil microbes. Gritty powders,

[67] *The New Farm*, May/June 1984, p. 14–15.
[68] *Acres U. S. A.*, March 1983, p. 40–41.
[69] R. Wolf, *Organic Farming*, 1977, p. 272–276.

mainly diatomaceous earth, are a non-toxic way to kill insects. When dusted on plants, insects or mites pick up the grit and die by drying out when their protective cuticle is scratched. Dusting an animal is a good cure for lice or fleas, and diatomaceous earth given orally (mixed with food) is an excellent de-wormer (about a handful per large animal per day for a week). Diatomaceous earth can be purchased economically from swimming pool supply companies.

Insect traps are being used successfully in certain situations. Light and electrocution traps attract night-flying moths, such as corn borers, corn earworms, cutworms, and armyworms. Pheromone (sex-attractant odors) traps are used to help control fruit flies and Japanese beetles. Or the pheromones are sprayed in the environment to confuse the insects and reduce mating effectiveness. These methods reduce pest numbers somewhat but may not provide effective control.

Natural insecticides (called *botanicals*) made from certain plants are comparatively safe for the environment. They break down quickly. Botanical insecticides currently available include pyrethrum (or pyrethrins), rotenone (or derris, toxic to fish; do not use near water), sabadilla, quassia, hellebore, nicotine (from tobacco), and ryania. They would be first choice if your pest problem requires immediate killing of pests.

The long-used insecticides called the *inorganics* include some that are relatively safe ecologically (sulfur, lime-sulfur), but others that are very toxic and should be avoided (arsenic and mercury compounds). Copper sulfate is an effective fungicide and supplies the trace element copper, but can kill soil life in large amounts.

The least desirable insecticides are the ones most used today, the *synthetic organics*. They have been widely used since World War II and have led to serious ecological problems. Some are slow to break down, while others are rapidly degraded. Several have already been banned or restricted because of their pollution or cancer-causing potential. Pesticides banned or restricted for general use in the U. S. include DDT, DDD, chlordane, lindane, malathion, parathion, aldrin, dieldrin, endrin, heptachlor, fluoroacetate (compound 1080), thallium sulfate, 2,4,5-T (liquid), demetron (Systox), acrylonitrile, aluminum phosphide (Phostoxin), methyl bromide, ethylene dibromide, strychnines, cyanides, azinphosmethyl (Guthion), and methomyl (Lannate, Nudrin).

Many pesticides are very toxic to honeybees (including carbaryl (Sevin), parathion, diazinon, ethion, cygon, methomyl, Rogor, Imidan, Dibrom, azinphosmethyl, Furadan, Penncap, Spectracide, Dursban, malathion, and phosdrin). Pesticides that are relatively safe for honeybees include methoxychlor, Fundal, most miticides (acaricides), most fungicides, and most herbicides. Bees are more valuable to us for their pollination services than for the honey they produce. Billions of dollars worth of fruit, vegetable, and forage crops depend on insect pollinators.[70]

If you must use a synthetic organic insecticide to save a crop, go ahead, but if possible reduce the dosage. The use of a surfactant (wetting agent) in the spray can allow you to reduce the amount of pesticide considerably. Always follow label precautions and instructions, and NEVER increase the dosage on the misguided philosophy that "if that much is good, twice as much is better."

Natural weed control. We have seen earlier in this chapter that most weeds grow best in out-of-balance soil with poor structure. So obviously, the number one priority in solving a weed problem is to correct whatever is wrong with the soil. A soil test can spot nutrient imbalances, but a shovel is your most useful diagnostic tool. By simply digging several holes in your field you can determine the general tilth (by ease of digging), spot waterlogging, a hardpan, and anaerobic soil (by how far down small feeder roots extend; do buried manure and plant residues decay rapidly?).

But until soil problems are corrected you may have some weed problems to contend with. As with pests, try to avoid use of synthetic chemical herbicides. If they are needed, using a surfactant (wetting agent) may allow you to reduce dosage. Here are some natural methods:

1. *Seedbed preparation.* For row crops, if you prepare an ideal seedbed in the row but leave the soil rough and cloddy between the rows, the crop should sprout first and get a head start on weeds. Or you can cultivate and delay planting, allow the weeds to sprout, and then till them out as you prepare the seedbed. Use of a ridge-planting system along with cultivating and mowing can effectively control weeds.[71] Vigorous crop plants will shade out weeds and are better able to withstand weed competition. Weeds that appear later in the season decrease yields only slightly.

2. *Mechanical control.* Timely cultivation with shovel, rotary, or disk cultivators, rotary hoes, and spike tooth and spring tooth harrows is an effective weed control method. Get the weeds as early as possible, before their roots go deep. Cultivation is also beneficial in loosening and aerating the soil, and in dry weather, the "dust mulch" it leaves discourages weed seed germination. Be careful not to prune the roots of the crop plants. Hilling of soybeans and especially corn not only buries

[70] *Acres U. S. A.*, July 1982, p. 19.
[71] *The New Farm*, July/August 1984, p. 18–20.

weeds but also stimulates corn prop root growth. Frequent mowing of weeds will weaken them by depriving them of food from the leaves. This is useful for already established weeds. If at all possible, prevent weeds from going to seed.

3. *Cover crops.* Planting a leafy or dense cover crop will shade out most weeds. Small grains (wheat, barley, rye) and buckwheat work well and can also later be turned under or mowed as green manure. Forages such as alfalfa or clover are excellent and help build the soil. Grasses and wood-chip mulches work well in orchards. Planting the shade crop early will allow it to get a head start on the weeds. Intercropping, planting a different crop between the rows of a row crop (such as soybeans between corn), helps control weeds, but the yield of the low-growing crop will be decreased. Although they are not cover crops, corn, buckwheat, or soybeans can shade out non-grassy annual weeds. If a crop species is grown in a climate for which it is not well adapted, it will have to struggle and will not compete well against weeds. Planting varieties that are adapted to your area is best.

4. *Allelopathy.* One reason some of the above cover crops control weeds is not only by shading them out, but also allelopathy, the release of toxic substances (phytotoxins) that kill or inhibit other species of plants. The phytotoxins can be released from the roots of a growing plant or can be produced when plant residues decompose in or on the soil, sometimes being released by certain soil microbes.

Both weeds and some crop plants can release phytotoxins. That is one reason it is good to get a crop off to a head start. In fact, phytotoxins from plant residues can injure the next crop in a rotation sequence. In a University of Nebraska study, stubble-mulched wheat residues reduced corn germination by 85% *in years of normal or above normal precipitation* (this suggests improper, anaerobic decomposition to me). In descending order, alfalfa, birdsfoot trefoil, ladino clover, red clover, reed canary grass, brome grass, timothy, and orchard grass residues were found to release phytotoxins. Rye, fescue, sunflowers, oats, and sorghum have been tested and found to control weeds, some over twice as powerfully as 2, 4 - D herbicide. Some farmers and orchardists are saving from $20-38 per acre using allelopathy instead of herbicides. [72]

5. *Natural enemies.* Some weeds have been successfully controlled by turning loose insects that eat them. Examples include a leaf beetle to control Klamath weed (St. John's wort), a flea beetle to control alligator weed, a moth to control three European ragweeds, a moth and a flea beetle to control tansy ragwort, a moth to control purple nutsedge, seed-

[72] *The New Farm*, February 1984, p. 34-35; November/December 1983, p. 6.

eating weevils to control puncturevine, and a weevil to control musk thistle.[73]

Naturally-occurring fungus diseases are being developed to control certain weeds in soybeans, corn, potatoes, rice, tobacco, and citrus. Some are already available commercially, and they are sometimes called "mycoherbicides." Weeds controllable by mycoherbicides include velvetleaf (buttonweed), sicklepod, leafy spurge, prickly sida, northern jointvetch, and citrus milkweed, some of which are resistant to herbicides. The natural control is thus more effective and cheaper than herbcides.[74]

Grazing animals (sheep, goats, cattle, horses, geese, and chickens) have been used to control orchard weeds in California.[75]

Corn stalk rot—why? To close this chapter on what *really* causes diseases, weeds, and pests, let's look at one specific disease problem which is increasingly common—stalk rot of corn. Today it is easy to drive around in the corn belt in autumn and see one field after another with stalks broken and lying at crazy angles—a headache to harvest and a terrific loss of yield, from 3 to 18%, according to M. E. Michaelson.[76]

The experts say stalk rot and its companion, root rot, are "caused" by fungi—*Diplodia*, *Gibberella*, *Macrophomina*, and *Fusarium* to be exact. But is fungal invasion of roots and then the stalk the *real* cause? The fungi happen to be common in the soil.[77] Not all plants and not all fields are affected; the disease is worse in some years. What conditions allow the fungi to invade? Let's put together a jigsaw puzzle picture of what really causes corn stalk rot by quoting relevant passages from scientific research reports, available in most large libraries. I have added some translations into everyday language in brackets and emphasized key words by adding **boldface** lettering.

1. "Corn plants . . . approaching maturity are ordinarily attacked rather than young, **physiologically active** [actively growing and metabolizing] plants."[78]

2. "The spread of *D.* [*Diplodia*] *maydis* in the corn plant is influenced by the **physiological state** [state of health] of the invaded cells."[79]

[73]G. C. Kingman, F. M. Ashton & L. J. Noordhoff, *Weed Science*, 2nd ed., 1982, p. 26; D. J. Borror, *et al.*, *ibid.*, p. 11.

[74]*The New Farm*, November/December 1982, p. 13; *Wisconsin Agriculturist*, January 26, 1980, p. 44; *Farm Industy News*, March 1982, p. 47.

[75]*Acres U. S. A.*, March 1982, p. 16–17; September 1984, p. 16–18.

[76]*Phytopathology*, 1957, vol. 47, p. 499.

[77]J. L. Dodd, *Plant Disease*, 1980, vol. 64, p. 533.

[78]A. J. Pappelis, *Phytopathology*, 1965, vol. 55, p. 623.

[79]D. S. Wysong & A. L. Hooker, *Phytopathology*, 1966, vol. 56, p. 34.

3. "Histological observations [examination of tissues under the microscope] indicated that the spread of *D.* [*Diplodia*] *zeae* was **limited to dead cells**. Spread of *Diplodia* was **inhibited by living cells**."[80]

4. "Pappelis correlated resistance to stalk rot from *D.* [*Diplodia*] *zeae* with the density of pith tissue in the stalk. Pith tissue of high density had a high percentage of living cells, whereas dry fluffy pith of low density was composed mostly of dead cells."[81]

5. ". . . root and stalk rot is preceded by a loss of cell turgor [internal "water pressure;" correlated with health of cell] and a **reduction in sugar levels** in the pith, whereas resistance to stalk rot is associated with the maintenance of succulent stalks and **high sugar levels** until after physiological maturity."[82]

6. "Resistance appears to depend upon the maintenance of a certain degree of **physiological vigor** [good health] . . . Conversely, susceptibility is due to a **lack of vigor** as the plant matures. Physiological vigor depends on a steady respiratory rate supported by a continuous supply of carbohydrate reserve, always assuming an **adequate and balanced level of nutrients**. When the vigor of a plant drops below a certain level because of **stress conditions** under which it is grown it becomes susceptible to invasion by certain saprophytes [plants living off of dead organic matter] and weak parasites."[83]

7. ". . . **environmental stresses** reduce photosynthetic [food making] capabilities of host plants . . ."[84]

8. ". . . root and lower stalk tissues are decayed by several microorganisms as the tissues lose their **metabolically dependent defense system** . . ."[85]

9. "The roots were the site of initial infection, and the pathogen progressed into the stalk by way of infected roots."[86]

10. ". . . the greater the infestation of the European corn borer, the greater the amount of stalk rot. The holes made by the insect provide easy avenues of entrance for fungi into stalks, shanks, and ears."[87]

11. "An **antifungal** phenolic compound [a natural disease-fighting chemical] has been isolated from corn roots and stalks . . ."[88] ". . . ex-

[80]A. J. Pappels & F. G. Smith, *Phytopathology*, 1963, vol. 53, p. 1100.
[81]J. Craig & A. L. Hooker, *Phytopathology*, 1961, vol. 51, p. 376–377.
[82]R. E. Wall & C. G. Mortimore, *Canadian Jour. of Botany*, 1965, vol. 43, p. 1277.
[83]C. G. Mortimore & G. M. Ward, *Canadian Jour. of Plant Science*, 1964, vol. 44, p. 451.
[84]J. L. Dodd, *Phytopathology*, 1980, vol. 70, p. 534.
[85]J. L. Dodd, *Plant Disease*, 1980, vol. 64, p. 534.
[86]J. Craig & A. L. Hooker, *ibid.*, p. 382.
[87]J. J. Christensen & C. L. Schneider, *Phytopathology*, 1950, vol. 40, p. 284–289.
[88]C. G. Mortimore & G. M. Ward, *ibid.*

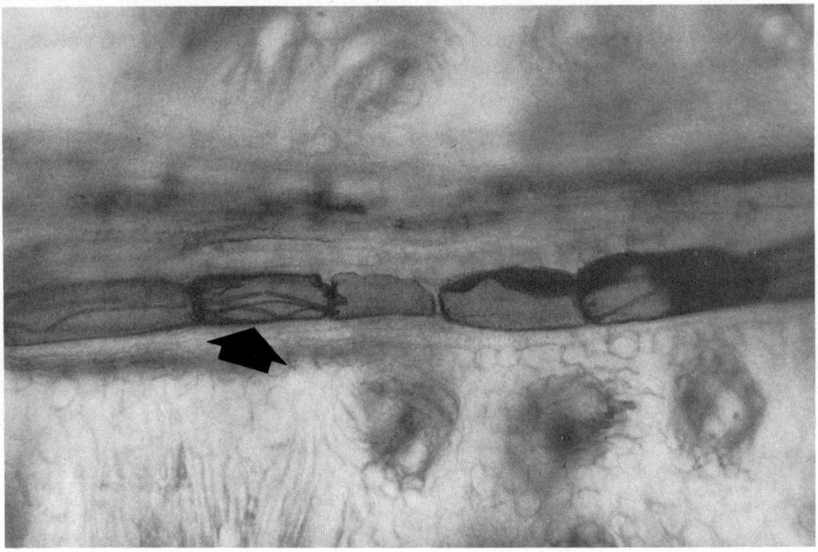

Invading corn stalk rot fungi in base of stalk. Magnified 100 times by microscope.

tracts from young corn stalks inhibit growth of two stalk-rotting organisms, *Fusarium moniliforme* Sheld. and *Gibberella zeae* (Schw.) Petch. Loomis *et al.* found **resistance factors** in corn which were effective against the European corn borer . . ."[89] "The occurrence of a cyclic hydroxamate . . . (DIMBOA), in maize [corn] was first reported in 1959 . . . Beck & Smissman . . . found that it inhibited insects, fungi and bacteria."[90] "It is suggested that this compound is equally able to play a role in the resistance of completely developed plants of maize [corn] to injury of borer larvae . . . and of other fungi, like *Fusarium* responsible for stalk rots."[91]

12. "A high nitrogen to potassium ratio greatly increased the severity of internal rot, the amount of stalk breakage, and premature dying. One method to reduce the severity of basal stalk rot or nodal rot may be through the prevention of early death of nodal cells by controlling soil fertility . . ." ". . . **unbalanced soil fertility always was conducive to stalk rot of corn**. . ."[92]

[89]N. J. Whitney & C. G. Mortimore, *Nature*, 1959, vol. 183, p. 341.
[90]R. M. Couture, D. G. Routley & G. M. Dunn, *Physiological Plant Pathology*, 1971, vol. 1, p. 516.
[91]P. M. Molot & P. Anglade, *Annuales des Epiphyties*, 1968, vol. 19, p. 94.
[92]A. J. Pappelis & L. V. Boone, *Phytopathology*, 1966, vol. 56, p. 850, 852.

13. "An increase in the concentration of phosphorus . . . reduced the development of both stalk- and root-rot of corn . . . the mineral balance most consistent with good growth of corn and inhibition of stalk- and root-rot is a medium level of nitrogen and potassium and a high level of phosphorus."[93]

14. "Potassium chloride [muriate of potash], applied at . . . 150 lb/acre potassium, decreased natural root necrosis [decay] and delayed death of corn plants, but **increased stalk rot** when the stalks were inoculated with *Diplodia maydis* . . . rates of [potassium chloride] of 101 kg/ha (90 lb/acre) of chloride caused a depression in plant growth, delayed maturity, and decreased yields."[94] Ammonium chloride treatments were included on selected plots . . . to determine whether or not reductions in lodging . . . were due to chloride or potassium in the KCl [potassium chloride]. It is concluded that the reduction in lodging was due to potassium and not chloride."[95]

15. "Rate of photosynthesis in each plant is affected by soil microenvironmental [small-scale] factors, leaf damage, and plant competition; but larger macroenvironmental [large-scale] **stresses** such as cloudiness, drought, **fertilizer imbalance**, **poor soil waterholding capacity**, and leaf disease epidemics probably are more important."[96]

Summary. In case you got lost, let's summarize. Corn root and stalk rots develop in plants that are not vigorous, with a low sugar content (low rate of photosynthesis), and dying pith cells. Stresses from such things as drought, cloudy weather, high stand populations, leaf damage, and insect attack allow the fungi to get started. But I would trace the real cause back to the soil: **nutrient inbalance** and **poor soil structure**. Healthy plants produce antifungal chemicals, but those under stress apparently do not.

Now that we have seen some of the main causes for our problems, let's look at ways to solve them, to rescue sick soil and get started on the road back to healthy soil.

[93]P. Thayer & L. E. Williams, *Phytopathology*, 1960, vol. 50, p. 212, 214.
[94]J. W. Martens & D. C. Arny, *Agronomy Journal*, 1967, vol. 59, p. 332, 499–500.
[95]W. C. Liebhardt & R. D. Munson, *Agronomy Journal*, 1976, vol. 68, p. 425.
[96]J. L. Dodd, *Phytopathology*, 1980, vol. 70, p. 535.

CHAPTER 5

A Better Way

"Be thou diligent to know the state of thy flocks, and look well to thy herds . . . He that tilleth his land shall have plenty of bread; but he that followeth after vain persons shall have poverty enough." Proverbs 27:23; 28:19

Perhaps you are convinced that natural farming methods are the way to go, but you don't know how to switch over. Or you are worried about lower yields or insect infestations if you don't use the methods that you and your neighbors have been using. These doubts and questions are understandable and common.

You know things are bad and getting worse, and you are looking for a way out of the vicious circle you are in—higher expenses and lower income—harder soil and lower crop quality—more weeds and bugs—sick animals. These are symptoms of problems. Problems that trace back to sick soil.

But they are problems that can be overcome. You need to (a) recognize the problem, (b) find the cause or causes, and (c) eliminate or overcome or prevent the cause(s).

Most sick agricultural soil wasn't that way before it was tilled. Virgin soil is usually pretty fertile and has good tilth. It is from unwise use (or abuse) by man that soil has deteriorated. It probably took years, even decades, for the damage to be done. It may take a number of years to make much progress in some cases, while in others amazing changes can be made in one year. So, to correct problems in the soil, the basic approach to use is to (a) stop doing wrong things, and (b) start doing right things.

Restoring sick soil. Let's look at some practical methods for restoring soil fertility and good tilth to abused soil. We will first look at methods that should work on most types of soils and then later we will cover methods of dealing with some problem soils, such as acid soils, alkaline and dryland soils, waterlogged soils, and soils polluted by toxic substances.

Problem analysis. The first thing a doctor does before treating a sick patient is diagnose the illness—use his trained eye, knowledge, and perhaps some lab tests to find out what's wrong.

So before rushing out blindly to correct your sick soil, you had better have a fairly good idea of what is wrong with it so you can do a more effective job and keep from making costly mistakes.

If you feel inadequate to diagnose your soil's ills, you may want to hire a professional consultant, although it can be risky if he also sells a product. His product may help, but you might make faster progress correcting sick soil if he can recommend a variety of materials, depending on need. Actually, for many if not most sick soils, if you have a little background knowledge of how soil and soil life function to nourish a plant, and if you have an observant eye and a lot of common sense, you can probably correct 90% of your soil's problems.

Call a spade a spade. The first and most important "scientific instrument" you should use to test your soil is a shovel! A long, narrow spade is best. Walk around in your fields and observe your soil carefully. Does it feel spongy when you walk on it, or does it more closely resemble a parking lot? Does it crust when dry or stick to your shoes when wet?

When you attempt to thrust in the spade, does it sink in at least six or eight inches with little effort, or do you have to jump on the spade to get it in two inches? When you pull the spade handle back, does the soil crumble into loose clumps or does it fracture into hard clods? See any earthworms?

Try digging deeper than the "plow layer." Do you hit a dense layer—a hardpan? Dig down through it and see if the soil is looser below it. In wet weather, the soil above a hardpan may be waterlogged, while below it is bone-dry.

Are there crop residues such as corn stalks several inches down that are a year or more old and still not decayed to humus? Either the soil is anaerobic or the life has been killed by toxins.

Take a lump of topsoil in your hand. Is it easy to crumble? When you squeeze some soil in your fingers, does it become a compact wad of clay? Smell your soil. Does the topsoil have that rich, pleasant odor you associate with humus or a fertile garden soil? If so, beneficial actinomycetes are alive and well (see Chapter 2). If soil (especially subsoil) has a foul, putrid odor, anaerobic fermentation is occurring, probably with toxins being produced.

Compare these observations in soil from good fields and problem fields or problem spots in a field. You can tell an awful lot about your soil when you get off your tractor and "get down to earth."

If you have read Chapters 2–4, you can probably analyze some of the most common soil problems: low humus, sterile, tight, compacted, anaerobic soil. But there's more.

What plants tell. Dig up the root system of a plant, such as corn or soybeans, or even pigweed. How far down are there a lot of fine feeder roots (taproots may extend deeper)? This shows the depth of your soil's aerobic (aerated) zone. Do your soybean, alfalfa, clover, snapbean, or other legume roots have the root nodules that they are supposed to? Nitrogen-fixing bacteria need well-aerated soil to form nodules. If there are nodules, cut some open. Are they pink inside or a sickly yellow or greenish? Only pink nodules are doing their job. Soil must be free of toxins in order for nodules to grow and function properly.

If your soil is plagued by weeds, check the box in Chapter 4 to see if they indicate a particular soil problem. But don't forget that a few kinds of weeds indicate good soil.

Are the plants growing in your soil healthy looking? Have yields or quality been poor in past years? A lot of bugs or diseases? These are more indications of sick soil—perhaps anaerobic or toxic.

Deficiencies of mineral elements can cause visible symptoms in the leaves of plants, but their diagnostic value is limited. They usually show up only when the deficiency is severe, and by the time you see them, it may be too late to correct the problem that year. Also, there may be more than one deficiency at once, with diagnosis being difficult. Nevertheless, the accompanying box lists the most common mineral deficiency symptoms that apply to most crops (a few crops may react differently).

The "deficiency" of an element may be due to different causes. There may be a true deficiency or lack of the element in the soil. Or there may be plenty of the element, but it simply is not available to the plant in sufficient amounts. This can be caused by high or low pH (see diagram in Chapter 4, on p. 76) or by an *imbalance* (too much or too little) *of another element* (see table in Chapter 4, on p. 75). And then there can be plenty of an available element, but if the plant is stressed or sick, the roots may not be able to absorb it. Stresses can come from drought, flooding, anaerobic soil, cloudy and cool weather, and toxins in the soil (see Chapter 3).

So again, plant deficiency symptoms have their limitations in diagnosing soil problems. For similar reasons, the more sophisticated plant *tissue testing* is also limited in usefulness. It tells you—very accurately—how much of each element is present in the plant, but cannot tell what is wrong in the soil.

Soil tests. Another tool to help diagnose soil problems is soil testing. However, many farmers and advice-givers assume that soil tests are super-accurate and that you can tell exactly how many pounds per acre of fertilizer you need to add this year.

The truth is, there are a whole lot of variables to consider. First, there are many different methods of testing soil and different philoso-

General Mineral Deficiency Symptoms of Plants

(compiled from various sources)

- **nitrogen**—slow growth of top and roots; leaves turn yellow-green when young to orange, red or purple when mature (veins may turn purple), beginning from bottom leaves to top; reduction in number of flowers and yield of grain or fruit; slow growth and delay of opening of buds.
- **phosphorus**—similar to nitrogen deficiency but leaf color either dull bluish-green with tints of purple instead of yellow or red, OR leaves dull bronze colored with purple or brown spots (leaf edges often brown, especially in potatoes).
- **potassium**—if mild deficiency, stems thin; if severe, stems stunted or die; leaves usually dull bluish-green, often yellow streaks between veins, followed by browning of tips and edges, and development of brown spots near edges; leaves rolled (begins at lower leaves); poor root growth; poor development of flowers and grain or fruit.
- **calcium**—begins at upper leaves: leaves very distorted and curled at edges; edges appear ragged and leaves may have thin yellow bands or be brown, then die; roots poorly developed.
- **magnesium**—leaves turn yellow (sometimes between veins, sometimes in spots or streaks), then turn brown and die (starting at bottom of plant); grain or fruits poorly developed.
- **manganese**—similar to magnesium, but starts at top of plant.
- **sulfur**—reduction of growth; curling down of leaves at tips; similar to nitrogen deficiency except that lower leaves are not lost.
- **boron**—begins at upper leaves: leaves become light green (lighter at base, veins turn purple, leaves may have faint streaks and yellowing, then turn orange) and stop growing, roll down at tip, with tissue breakdown at base; poor growth of tops and roots; flower buds drop off; no grain or fruit.
- **iron**—severe yellowing of newer leaves (either spotted or total); more often visible in trees than yearly crops.
- **zinc**—mottled spots on leaves, first yellowish, then yellow or purple-red (appears late in summer, then leaves drop early), leaves become crinkled and small.
- **molybdenum**—similar to nitrogen deficiency, since molybdenum is necessary for nitrogen-fixing bacteria; leaves becoming yellow between veins, first on mature leaves, then to younger; young leaves may become severely twisted and eventually die.

phies of making recommendations based on test results. You can literally send identical samples to two (or more) labs and get two (or more) completely different sets of figures and recommendations back. The numbers aren't wrong; it's just that different ways of measuring something were used. Also, many labs recommend too high amounts of some fertilizers, just to "be on the safe side." Recent articles in *The New Farm* magazine have pointed out the variability of soil testing labs (summary article, February 1983, p. 21–25; also see February 1984, p. 22–23, July/August 1984, p. 21–23, and September/October 1984, p. 18–20).

Most labs use methods that test the soil under artificial (dried, finely ground) rather than field conditions. They generally determine the total amounts of some nutrients rather than what is readily available to plant roots at any one time. A more realistic method uses weak extracting fluids that simulate the nutrient-extracting power of roots. Most labs will run these soluble, readily available tests if requested.

Most farmers have their soil tested once every year or two at the most. But soil is constantly changing, and nutrients that may be unavailable in March or September may be in abundance in June, through microbial action. It is best to monitor soil fertility at least two or three times a year (spring, summer, and fall) to see what is going on and to spot developing problems. That way corrective measures may be made the same year and a crop saved.

Soil testing kits are available which allow the farmer to quickly get a rough idea of his soil nutrient levels and pH. If you build up a few years of soil testing records and compare results from good and poor fields, you will soon be able to interpret your tests yourself. Many of the professionals tend to recommend too much potassium and too little calcium and phosphorus.[1]

To collect soil samples, take a small amount of soil from the plow layer (upper 6–10 inches) from 10–30 typical areas of the field. Avoid field edges, fertilizer bands, or small problem areas (unless you specifically are sampling a problem area). Do not touch the samples with your fingers or rusty iron tools. You can dig with a shovel and use a wooden instrument to get a few lumps of soil into a paper or plastic bag. A total of 1/2 to 1 pound (1 to 2 cups) of soil from a field is all that is needed. Label each sample with a number or letter, and with your name and address if you are sending it to a lab.

Can't always work miracles. Every field is different, climate varies, and different crops have different requirements. Therefore I cannot tell you exactly what to put on your fields to correct soil problems without knowing the details. In this book I can only give general principles and methods that should work in most cases, on most kinds of soil. But unusual conditions and soil problems do exist which may require the expertise of a reliable consultant. The methods and materials I mention

[1] R. M. Heinicke, *Acres U. S. A.*, November 1981, p. 1, 8.

should help when used properly, but I cannot guarantee total success without knowing the situation firsthand.

By knowing the basic principles of how soil, soil life, and plants function, you should be able to use common sense to decide what to do and to diagnose problems.

General guidelines for soil correction. Following are some general guidelines for correcting abused soil. They should work for most types of soil under moderate climatic conditions, such as would exist in the central and eastern United States. In the dry areas of the West and Southwest (without irrigation), other problems may occur, such as alkaline and saline soils. Guidelines for these and other types of problem soils will be given later in the chapter.

In these General Guidelines, I am assuming that your soil may have such problems as low humus, little soil life, compaction, low or unbalanced fertility, weeds, hardpan, waterlogging, and toxicity from pesticides or wrong fertilizers. So where do you begin, with any or all of those problems? What do you do first?

1. *Analyze the problems.* The first thing to do is to figure out what is wrong. Follow the pointers given just previously in this chapter. Walk in your fields, dig, and observe the symptoms. Make a list. Have your soil tested, especially to spot serious deficiencies or imbalances.

2. *Make plans.* Decide on your short and long-range "plan of attack." If you have relied heavily on chemical pesticides and want to switch to natural "non-chemical" methods, it is best to do so gradually, over anywhere from three to eight years, and to work intensively with smaller fields (up to 30–50 acres) on about 20–25% of your farm. Of course, owned land should have priority over rented land.

If you have livestock or other animals that you raise the feed for, be sure to restore those fields first, and leave cropland that you raise cash crops on for later. Don't forget to put extra effort into correcting your garden, since you and your family are most important.

If you are rotating crops, use a rotation that will do the most to improve soil and control weeds. Eventually, when the soil is in good condition, you may not even need to rotate.

Figure out your budget: how much you will spend on fertilizers and soil amendments, extra fuel for cultivating weeds, and how much you will save by not using pesticides. Allow some leeway for emergency pest control the first couple years.

Get in mind what you are trying to do. Depending on what your exact problems are, you will probably be (1) *improving soil structure* (tilth), loosening compact soil in order to allow air in, which will (2) *increase beneficial soil organisms*, which will (3) *increase humus content* and (4) *improve plant health* and resistance to diseases and pests; while at the same time, (5) *correcting soil nutrient imbalances* and (6) *detoxifying soil*, which will help in all of the above, as well as (7) *reducing weeds*.

Don't expect to do all of that the first year. It will probably take several years to show great improvement, but once you stop doing harmful things and start doing helpful things, rejuvenation will begin and improvement should be noticed the first year. Keep good records of all that you do: materials applied and amounts, tillage, crops and varieties grown, weather, and results.

3. *Improve soil structure.* There are several additives and methods that can improve soil structure. Details on the materials are given in Chapter 7. Some methods and materials are definitely more expensive than others.

A. *Rock fertilizers* can loosen soil. Soft rock phosphate (colloidal phosphate) and high calcium lime do this, plus supplying phosphorus, calcium, and trace elements. These two work best together, at a ratio of two parts lime to one part soft rock phosphate, unless soil is already high in calcium (if applied separately, apply the soft rock phosphate first). Certain other commercial rock fertilizer products also help loosen soil and supply minerals (see Chapter 7).

B. *Organic matter*, whether fresh or as humus, will loosen soil. Humus or compost is much preferred, because if too much fresh organic matter (green manures or animal manures, crop residues) is incorporated in tight soil, it may turn anaerobic and release toxins, worsening the problem. However, coarse crop residues (corn stalks, stubble), if partly incorporated and partly exposed, will help air to enter the soil.

Animal manure sources are preferred, since they are richer in nitrogen than plant matter, but tests have shown that a mixture of animal manure and plant matter gives better results, and a mixture of manure and rock fertilizers works even better. If animal manures are not available, animal wastes such as tankage, blood meal, fish meal, etc., or high nitrogen plant matter, such as spoiled alfalfa or clover hay, are good (use small amounts at first). Sewage sludge, cannery wastes, old straw, city grass clippings and leaves, and peat are other excellent sources of organic matter.

If the organic matter decomposes aerobically, it will supply most of the soil nutrients needed by plants. But if a lot of fresh organic matter is applied too soon before planting, its decomposition will temporarily tie up nutrients and starve the crop. If the organic matter is low in nitrogen, microbes will temporarily "rob" nitrogen from the soil. Fresh organic matter is best applied in the fall in northern climates, or at least a few weeks before planting in the spring. Compost has the advantage of being already decomposed, but is still rich in nutrients. Compost helps sandy soils the most; fresh organic matter helps tight clay soils more than compost. Compost raises the soil's organic matter content over the long run much more than fresh manure.[2] See Appendix B for details of composting.

C. *Cover crops and sod* are one of the best ways to loosen soil and add organic matter. Perennial grasses, with their multitude of fine roots, penetrate tight soil and add much humus as the roots gradually die. These include such species as smooth brome, orchard grass, fescues, bluegrass, timothy, and reed canary grass. Legumes (alfalfa, clovers, vetches, etc.) are excellent since they add nitrogen if soil conditions allow root nodules to form. It is best to let the cover crop grow for more than one year. If you do not need the hay, cutting and letting the plant matter lie and decompose adds greatly to soil humus (or do so with the first cutting of alfalfa and harvest the rest).

If the soil is so poor that the above crops will not grow well, just let it go to weeds for one or two years. Deep-rooted annual weeds break up a hardpan, aerate soil, and bring up minerals from the subsoil. Common deep-rooted (with a taproot) weeds include pigweed, goldenrod, ragweed, sunflowers, thistles, nettles, lamb's quarters, horseweed, wild lettuce, milkweeds, smartweeds, ironweed, dandelion, and wild sweet clover. Shallow-rooted or grassy weeds are not as valuable, but do cover the soil and prevent erosion, plus can be plowed under as green manure. Weeds should be mowed several times, or at least before they go to seed.

Legumes and annual weeds do take a lot of moisture out of the soil, which can be harmful in dry climates. Perennial grasses do not use as much.

When any of the above cover crops are plowed under, do so as late as possible in their development, but while they are still green.

D. *Humates* and humic substances, brown coal, and peat are something like "preserved humus," and often have beneficial effects on soil structure, loosening and aerating soil, as well as stimulating microbe and plant growth.

E. *Seaweed* products, and live *algae* inoculants, can improve soil structure by releasing glue-like substances. A healthy population of *soil bacteria* and other microbes also do the same thing for free. They need food to live on, namely fresh organic matter. Commercial bacterial inoculants are not always successful in establishing beneficial organisms, either because of competition from existing soil microbes, toxic substances, wrong pH, or lack of food (organic matter). The best way to

[2] P. Ott, p. 327–334 in *Basic Technics in Ecological Farming*, 1982, S. Hill & P. Ott, eds.

establish beneficial soil microbes is to add a lot of organic matter (animal and/or green manures, crop residues, compost), but not too deeply in the soil (not over 5–6 inches deep). *Enzyme* sprays have also been found to improve soil structure, perhaps by stimulating soil organisms.

Earthworms do a tremendous job in loosening, aerating, and enriching soil. They can break up a hardpan. They do need food (organic matter) and freedom from toxins. They will generally come into an area naturally when conditions are favorable for them.

F. *Wetting agents* or surfactants often do a quick job of making soil more porous and improving drainage. But in other cases results are disappointing, or if soil has little nutrient-holding capacity (sandy soils), nutrients can leach away. See Chapter 7. Sometimes surfactants can alter soil structure and reduce available living space among soil particles for microorganisms.[3]

G. *Mechanical aeration and subsoiling* are quick ways of loosening tight soil and breaking up a hardpan, but if nothing else is done to improve soil structure, the soil soon settles back to its original condition. They would be good to use in very severe cases *along with* organic matter or other methods mentioned above. Rotary tilling organic matter into the upper several inches is an excellent practice, especially for poorly-drained soil. Experimental methods of injecting manure into the crevices made by a subsoiler have been successful in loosening tight soil.

4. *Correct major pH and nutrient imbalances.* At the same time that you begin improving soil structure, if soil tests or plant symptoms indicate seriously out-of-balance nutrients or *much* too high or low pH, these should be dealt with immediately.

Ideally for most areas and most crops, pH should be between 6.0 and 6.8, but in good soil with high humus and beneficial organisms, crops can grow well within a larger range of pH, say between 5.5 and 8.0. High humus and beneficial organisms tend to correct pH problems. Also, as we mentioned in Chapter 2, pH can change greatly in a growing season (in the midwest, it often drops during the summer), so a single pH test reading shouldn't be taken too seriously. But an extremely high or low pH should be corrected.

In average soils, high calcium lime is generally the best material to raise a low pH (acid), since it also supplies calcium, which is usually needed anyway. Dolomitic lime should only be used if the soil is very deficient in magnesium; it is a very hard stone and slow to become available. It may not be available for a year or 18 months. The finer the lime is ground, the more quickly it will act.

[3] *Acres U. S. A.*, August 1984, p. 16.

High pH (alkaline) is seldom a problem except in the West. These problem soils will be covered later.

Unless you know your soil has a serious trace element shortage, don't worry about the minor elements until you have the major elements in balance. The most important are calcium, phosphorus, potassium, and nitrogen. Generally, after these are in balance, there will be fewer trace element problems because humus and soil organisms make them available.

More details will be given on nutrients in the section on fertilization, later in this chapter.

5. *Build up humus.* As we have seen in previous chapters, adequate humus and the beneficial soil organisms that produce it are essential for healthy soil. Most soil organisms need oxygen, so well drained and aerated soil is required. But humus and soil organisms help make soil loose and well-aerated. So if you are starting at near zero, with "dead" tight soil, how do you start turning things around?

First, reduce or eliminate toxic materials that harm soil life and tighten soil (pesticides, anhydrous ammonia, high salt fertilizers, and excess magnesium on clay soil). Then, use one or more of the methods for improving soil structure previously covered in section **3**. Then, since adding fresh organic matter to tight soil can cause more problems, the best way to increase humus is to add compost. Any amount can be added at any time without harm, except that the nitrogen in compost may be too much for small grains if applied just before planting, causing lodging.

If compost is out of the question, fresh organic matter, in moderate amounts at first, should work, if it is kept in the upper several inches. A cover crop and green manure plowed or disked into the upper several inches will also get the ball rolling. Adding rock fertilizers along with the fresh organic matter helps greatly. There are sprays of beneficial microorganisms and/or enzymes that will speed decomposition of plant residues.

After tight soil begins to loosen up and the aerobic zone deepens to 5 or 6 inches, the amount of fresh organic matter can be increased (but not just before planting). Then, several years later, when the soil really is in good condition, you will need less, since soil organisms, air and water will supply most nutrients (soil life breaks down minerals and fixes nitrogen; rain brings nitrogen, sulfur, etc.). Of course some crops require more nutrients than others: corn needs a lot, small grains little. Still, the soil organisms do need some food (fresh organic matter) to live on, so some needs to be added yearly. Don't forget that plant roots, stubble, and stalks add much.

6. *Detoxify soil.* Generally, by the time you have done the above steps, any toxic residues will have already been eliminated. The following are excellent detoxifiers: (a) calcium (lime), (b) soil life (especially bacteria), (c) organic matter (humus, compost), (d) plowed-under green manure (chlorophyll), (e) humates, seaweed, enzymes, etc. (see Chapter 7), and (f) oxygen (keep soil well-aerated, with good tilth).

Most toxic residues will disappear in time, within three to five years.

7. *Control weeds, pests and diseases.* Again, if you have done the above things, your weed, pest and disease problems should already be lessening. Aerated soil with high humus and balanced nutrients should favor vigorous crop growth and discourage most weeds (especially keep the level of available potassium below available phosphorus). Healthy plants will resist pest and disease attack.

What weeds there are can be best controlled by cultivation, in row crops. Cultivation not only kills weeds but performs the valuable job of aerating soil, stimulating root activity and beneficial soil organisms.

If there is a weed, pest, or disease problem that is urgent and cannot be controlled by other methods (see Chapter 4), then the use of toxic herbicides, insecticides, fungicides, etc., can save the crop. If possible, use reduced rates of these dangerous chemicals. Adding a surfactant, or wetting agent, will increase effectiveness, allowing you to use less pesticide. Banding pesticides in the row will reduce the amount sprayed.

Guidelines for correcting problem soils. The above General Guidelines should work for most types of soils in moderate climates, but some special treatments may be needed for certain problem soils.

1. *Extremely acid soils; high organic matter soils.* Generally these two go together. In cool climates or where lack of aeration slows or prevents microbial action, organic matter builds up rather than changing into humus. Soils with 20–50% organic matter are called muck soils, while those with over 50% are called peat soils.

In other areas sandy soils can become depleted and acid. Over-use of ammonia, urea, and ammonium fertilizers can lead to too acid soil, probably with low organic matter.

Of course, one remedy for acid soils is liming. Liming materials contain a carbonate (CO_3^{--}) or hydroxide (OH^-) ion, which combine with hydrogen ions (H^+), which cause most of the acidity. The most commonly used liming materials are ground-up limestone, but there are two types: high calcium lime (calcium carbonate, $CaCO_3$) and dolomite (calcium magnesium carbonate, $CaCO_3 \cdot MgCO_3$). Dolomite is a hard rock and slow to go to work, plus excess magnesium in soil can lead to tightening and hardening. High calcium lime is preferred. Papermill sludge is an excellent source of high calcium lime, although it may be hard to spread.

But lime does more than reduce acidity. It loosens compact soil, stimulates microbial action, supplies calcium and some trace elements, and reduces toxins. Soft rock phosphate or basic slag would also help acid and clay soils.

In high organic matter soils, aeration of the soil by tillage and drainage (if waterlogged), speeds up the oxidation of organic matter, which will bring the soil into a better balance. In low organic matter acid soils, adding organic matter will reduce the acidity.

Reducing acidity will generally correct trace element deficiencies and/or toxicities that tend to accompany acid soils. Adding compost or manure will inoculate "dead" acid soils with soil organisms, even though they may already be high in organic matter.

2. *Alkaline and saline soils.* In the arid western U. S., including irrigated areas, high moisture evaporation concentrates salts of sodium, calcium, potassium, magnesium, and other elements. Also, very close to any ocean, low land can have too many salts. These soils have low availability of phosphorus, potassium, iron, and other trace elements. Soluble salts and/or certain elements in high amounts are toxic to plants. Soil tilth is usually poor and drainage often restricted, resulting in anaerobic conditions and crusting. Alkaline soils have high pH; saline soils have high salt levels.

Alkaline and saline soils must be well-aerated and drained to be corrected. Drainage allows excess salts to leach away. Subsoiling or deep plowing may help in some cases, but soil additives are also necessary. Applications of gypsum (calcium sulfate) or other sulfate-containing fertilizers produce acid to neutralize high-alkaline soil. Iron sulfate is expensive, but would be good if there is an iron deficiency. Even sulfuric acid (battery acid) can be sprayed to reduce alkalinity, but application of pure sulfur requires time-consuming (2–3 months) bacterial action to be effective. If an alkaline soil is low in calcium, a calcium material must be added (gypsum, lime).

Application of irrigation water after the addition of the above will speed up the action if there is good drainage.

Organic matter, preferably compost, or else sawdust, wood shavings, peat, or cottonseed meal, should definitely be added, in the upper layers of the soil, to improve soil structure, reduce toxicities, reduce alkalinity, and inoculate with soil organisms. If you make your own compost, add rock phosphate rather than lime. After soil starts to improve, fresh organic matter can also be added; cover crops and green manures are excellent. Growing a sod crop for a while will improve structure. If soil tests show iron, phosphorus, potassium, or other deficiencies, supplements may be needed until soil improves more.

Grow crops that are salt and alkaline tolerant, such as sugar and garden beets, cotton, barley, asparagus, tomato, carrot, potato, peas, lettuce, alfalfa, sweet clover, wheat, rye, sunflower, several grasses, etc. One of the best ways to correct saline soil is to grow alfalfa for four or five years. Plant row crops on ridges so salts will leach away from roots. Avoid irrigation with saline water. If ridges are used with irrigation, salts tend to accumulate on the highest part of the ridge during dry periods and can kill germinating seedlings. It is best to plant rows on the sides of the ridge. Keep the soil moisture level high while plants are growing, since this dilutes salts.

3. *Prairie and dry soils.* West of the Mississippi in the central U. S. and extending up into the Northwest (eastern Washington and Oregon), there are soils that receive minimal rainfall, are usually slightly alkaline and high in calcium, and (usually) low in humus. These areas are best suited for grazing and wheat or cotton, but with irrigation, other crops can be grown. Unfortunately, excessive and wasteful irrigation practices have seriously depleted the underground water supply in the Great Plains (the Ogallala aquifer). Unirrigated dryland farming should strive to conserve moisture. Crops such as corn and legumes, which require much water (thus irrigation) are not good investments in the long run. It is difficult to raise humus levels above a few percent, but the more organic matter there is, the better the soil will soak up and hold the little rain that falls. Humus will even absorb dew when there is no rainfall. Compost is the best way to add organic matter because it is already decomposed; fresh manure or plant residues may not be able to rot quickly in dry soil. The stubble mulching system (leaving most crop residues, such as wheat straw, on the surface) has been shown to give higher yields and leave more organic matter and nitrogen in the soil than conventional methods.[4]

Leaving the surface cloddy and/or with stubble exposed will aid in accumulating moisture (including snow) and reducing wind erosion. Periodical fallowing, generally with a stubble cover, is often used in dry areas to allow soil to recharge moisture and fertility levels. However, excessive growth of weeds results in moisture loss; cultivation with an undercutter can control them.

Proper fertilization of crops will actually conserve moisture, since

[4]*Acres U. S. A.*, August 1983, p. 20.

healthy plants produce greater yields per unit of water, although they may require slightly more total water.

4. *High rainfall soils.* In the southeastern U. S., along the Atlantic coast, and near the Pacific coast of Washington and Oregon, the soils have been weathered by high rainfall. Subtropical and tropical countries often have the same problems. High rainfall causes severe leaching of certain nutrients, plus erosion. Hot temperatures oxidize organic matter. Consequently, high rainfall soils tend to be acid and low in humus and nutrients.

What these soils need, in general, is organic matter and lime, plus any other nutrients or trace elements in which they may be deficient. These will need to be replenished frequently, because of the constant leaching. Humus will hold and make available a lot of nutrients, but in warm climates, fresh organic matter or compost must be frequently added.

In the meantime. While you are working with your soil, trying to get it back in shape, you have to make a living and raise adequate crops. The best approach is to do as much as your time and budget allow to correct soil problems, and in the meantime, use fertilization and management practices that are the most economical and in tune with nature. In-row supplementary fertilization allows you to grow excellent crops on less than ideal soil, economically.

Fertilization. Although in theory, humus and soil organisms can supply all the nutrients plants need, in practice soil conditions and weather aren't always ideal, leading to imbalances and shortages. Also, some crops require more nutrients than others, and during their period of peak needs, soil organisms may not be able to keep up an adequate supply if soil isn't in top-notch condition.

So in many cases, and while you are restoring abused soil, you will have to add supplementary fertilizers, either to supply what the soil is short of or to give the crop an extra boost at a critical time.

Which fertilizers are best? Which ones are harmful to soil and soil life, and which are not?

General principles. What do plants need to grow? The textbooks say some 17 elements make up plant tissue. Three of them—carbon, hydrogen, and oxygen—are gotten from water in the soil (H_2O) and the carbon dioxide (CO_2) in the air. Nitrogen-fixing bacteria and thunderstorms can supply some nitrogen from the air. These elements make up at least 98% of a plant, and they are mostly free (unless you irrigate).

The nitrogen used by plants can either be in the nitrate form (NO_3^-) or the ammonium form (NH_4^+). Nitrate is not tightly held by the soil and can easily leach into the groundwater, while ammonium is more tightly held. But in well-aerated soil, ammonium and ammonia are quickly converted into nitrate by beneficial bacteria (the process is called nitrification).

The nitrate is either used by plants or leaches away, so a continual supply of nitrogen is necessary throughout the growing season. Thunderstorms and nitrogen-fixing bacteria supply some, but a good supply of humus or supplementary fertilizer are the main sources for most crops.

In anaerobic or flooded soil, nitrates are converted by anaerobic bacteria to nitrogen gases which escape into the air (called denitrification). This is one of the many bad effects of anaerobic soil (see Chapter 3).

In many plants the nitrate form of nitrogen promotes rapid vegetative growth (stems and leaves), while ammonium nitrogen, if applied at the right stage of growth, causes the plant to "switch over" to producing flowers, seeds, or fruits.

The one or two percent of a plant's tissues that don't come from water and the air come from the mineral elements of the soil. Besides nitrogen (N—the chemical abbreviation), they include phosphorus (P), potassium (K) [those are called the major elements], calcium (Ca), magnesium (Mg), sulfur (S) [those are called the secondary elements], and the trace elements or micronutrients, which are needed in very small amounts: iron (Fe), copper (Cu), manganese (Mn), zinc (Zn), boron (B), cobalt (Co), molybdenum (Mo), and chlorine (Cl). Other trace elements may be added to the list as research continues. There isn't space now to mention the functions of all these nutrient elements in the plant; this is covered somewhat in my previous book, *The Rest of the Story. . . about Agriculture Today*, Chapters 4, 11, and 12.

Test first. As we have said earlier in this chapter, frequent soil tests should be made to see which nutrients your soil needs and in what amounts.

Which fertilizer materials? The 14 mineral nutrients needed by plants can be supplied by many different fertilizer materials. Some soil amendments, which improve soil structure or pH, also double as fertilizers.

Some materials farmers put on their soil are valuable fertilizers and help build up the soil, while other materials are highly toxic to plants and/or soil organisms if used the wrong way. Some are easy to apply and some are a pain in the neck. Some are instantly available to plants, while others may take months. Some are cheap per unit of nutrient, while others are expensive. You need to know what is going on your land if you want to still be in business in several years.

It is not easy to classify fertilizers because terms mean different things to different people. To a chemist, the word *organic* means any chemical substance containing carbon atoms, whether it is natural or man-made, and *inorganic* means non-carbon substances. But the original meanings of those terms are: substances produced by living organisms are organic and all other materials are inorganic. These original definitions are useful in referring to fertilizers.

Two other terms that can be confusing are *natural* (any substance found in nature, not produced by man) and *synthetic* (man-made, artificial). Some people use the word *chemical* to refer to synthetic fertilizers, but this is not good because *chemical* refers to any distinct kind of matter, whether natural or synthetic, such as water, table salt, or oxygen.

In our non-technical language, *organic fertilizers* include those derived from plants or animals, such as manures, sludge, bone meal, tankage, green manures, composts, and humic substances. We can divide the *inorganic fertilizers* into those that are natural (mineral deposits such as limestone, rock phosphate, potash, gypsum, greensand, granite, and fly ash) and those that are synthetic (such as anhydrous ammonia and the many "salt fertilizers," so called because they are made up of a positive and a negative ion). A few materials fit into more than one category, including muriate of potash (potassium chloride), which is mined from ancient salt deposits, and urea, which is found naturally in animal urine, but mainly synthetically manufactured.

Organics. In general, organic fertilizers contain complex substances (proteins, carbohydrates, fats, organic acids, vitamins, phenols, etc.) which are slowly broken down by soil organisms to release *both simple and complex* plant nutrients. Plants can absorb some complex molecules, such as amino acids, amines, phenol, and indole, in addition to the simple nutrients the textbooks all say they require. Thus, organic fertilizers provide plants with a *variety* of nutrients, not just one or two as do most synthetic materials. However, the organic nutrients are not all immediately available to plants, but this is good because it gives long-term fertilization. The complex organic molecules first serve as food for soil organisms, becoming available to plants when the organisms die; this counteracts leaching away of nutrients.

Inorganics. Most natural inorganics (rock fertilizers) also break down slowly, and they generally also supply a number of trace elements besides their main constituents. In contrast, synthetic or purified "salt" fertilizers are rapidly soluble and available to the plant and contain few trace elements. They give plants a "shot-in-the arm," a "force-feeding." But if soil is deficient in humus and beneficial organisms, it may be unbalanced nutrition for the plant. Also salt damage ("fertilizer burn") or stress can occur to plants, soil organisms can be killed, and soil structure harmed if too much or wrong kinds of synthetic fertilizers are used. See Appendix A for characteristics of commonly used fertilizers.

When to use what? The number of variables is too great to be able to spell out what specific fertilizer materials you should use. It depends on your soil type, soil test results, climate, crop growth, budget, etc.

If you can afford a soil building program, then broadcasting the right organic and rock fertilizers will give excellent results after a year or two (they are in general slow-release fertilizers). After the first, perhaps large, application, only small amounts may be needed in later years.

If your budget is tight, it would be best to give a full soil building treatment to a small acreage of a high priority crop (a high value cash crop or one that you feed to your animals), since a halfway application over a large acreage may not give visible results the first year. Then use the rest of your money on a maintenance program for your other acres. Row application, split applications, and foliar feeding can produce excellent crops a year at a time for a very low fertilizer cost per acre. Shopping around for the best buy, pooling orders with neighbors to receive bulk discounts, and doing your own blending are other ways to save money.

Any organic and rock fertilizers that you can get on the soil will help (provided they are what the soil needs and are used properly). More finely ground rock fertilizers will go to work faster, but may not last long. Then, if crops need supplementary nutrients, small amounts of certain synthetic fertilizers can fill in the gaps. A small row application (placed to one side or below the seed) can get seedlings off to a good start. If mid-season fertilization is needed, side dressing (perhaps split into two or three small applications) and/or foliar applications can make the difference between a moderate yield and a great crop. In side dressing, a small amount of fertilizer (perhaps 100 to 200 lbs. per acre) is applied between the rows on the surface. A convenient way to do this is to mount fertilizer boxes on your cultivator for dry fertilizers. Even dry fertilizers on the surface will dissolve from rain or dew and become available to roots just under the surface. Spraying of liquid fertilizers would also be effective.

Level, well drained loamy or clay soils, and soils with adequate organic matter respond the best to supplementary fertilization. Poorly drained, waterlogged, or drought-prone soils do not respond well. High sand content soils cannot hold nutrients well and require more fertilization (most economical in several small applications instead of one large dose, which can leach away). Very low (below 6.0) or high (above 7.5) pH reduces availability of several nutrients, especially phosphorus, except in very high organic matter soils (muck, peat), where pHs of 5.0–5.5 give best overall availability.

Synthetic fertilizers with a low "salt index" are best. The salt index is a measure of the fertilizer's ability to draw water out of living cells by osmosis. See the table of the salt indexes of common synthetic fertilizers in Chapter 3.

Fertilizer materials that cause a great change in pH should be avoided, or else applied long before planting, or placed far from the seed.

If supplementary nitrogen must be supplied, remember that nitrate nitrogen leaches easily, so another split application may have to be made after a heavy rain. Nitrate nitrogen promotes rapid vegetative growth (stems and leaves), so is desirable for hay or forage crops and early growth of seed crops.[5] Ammonium nitrogen applied (side dressed) when a seed crop is about halfway through its growing season will stimulate it to switch over to producing flowers and seeds or fruit. Obviously, you would not want to put an ammonium fertilizer on alfalfa or clover unless you were growing it for seed.

Most agronomists or fertilizer experts tend to recommend too much potassium in relation to the other major nutrients. Some potassium is necessary, but if phosphorus and calcium levels are too low compared to potassium, low quality, fibrous crops result, lacking high quality proteins and other nutrients (crops may grow well and look good, but quality is still poor).[6] The same thing is true of magnesium. Some is needed, but not too much in relation to calcium.[7] Calcium and phosphorus "build quality" into a crop. Unfortunately the "sufficient" levels of them on soil tests are usually too low.

Some synthetic fertilizer materials are much more harmful than others, even though they may be cheap and commonly used. Those materials to be avoided are:

1. *Muriate of potash* (potassium chloride, kalium potash, 0 - 0 - 60 or 62). Has the highest salt index of any common fertilizer material, and contains 47.6% chlorine. Some crops are highly sensitive to chloride (potato and tobacco), while others are moderately so (corn, soybeans, sweet potato, strawberry, cucumber, pepper, blueberry, citrus) in amounts over 100 parts per million chloride. Muriate of potash can injure or kill roots and interfere with the life activities of beneficial soil

[5]*American Vegetable Grower*, February 1982, p. 62.

[6]F. A. Gilbert, *Mineral Nutrition and the Balance of Life*, 1957, p. 54–55; D. Davidescu, p. 1095–1096 in *Fertilizers, Crop Quality and Economy*, 1974, V. H. Fernandez, ed.; W. A. Albrecht, *The Albrecht Papers*, 1975, p. 298.

[7]F. E. Bear (ed.), *Soils in Relation to Crop Growth*, 1965, p. 207.

microorganisms, including those that fix nitrogen. Lesser used chloride-containing fertilizers, such as calcium chloride and ammonium chloride would also cause problems.

2. *Ammonia* or other materials that release large amounts of ammonia (urea and diammonium phosphate). Ammonia gas is a cell poison. In large amounts in soil it can kill roots and soil organisms. Moderate amounts of ammonia in a liquid fertilizer (aqua ammonia) are not so bad. Ammonia is quickly dissolved in soil water and trapped on soil particles, and if soil bacteria were not killed, they convert it to nitrate nitrogen (the process is called nitrification).

Besides being toxic, large amounts of ammonia make humus soluble and easily leached. Thus over the years, humus is lost and the soil becomes hard.[8] In fact, soil injection of anhydrous ammonia was first developed during World War II to make soil hard to build landing strips for bombers in the South Pacific.[9] You can easily demonstrate how ammonia dissolves humus by putting a small amount of topsoil or compost in a glass of water. The humus will tend to float and the water will remain fairly clear. Add several drops of household ammonia, stir, and instantly the water will turn brown with dissolved humus.

So in general, the use of synthetic "salt" fertilizers is less desirable than using organic and rock fertilizers and soil amendments. The natural fertilizers are compatible with soil life and if used properly, build up and improve soil, while feeding plants a "balanced diet." Salt fertilizers, if used improperly, can kill soil life and roots, harm soil, and grow poor quality crops. On the other hand, some synthetic fertilizers, if used in small amounts, can be a needed supplement if a crop runs short of nutrients or can be used for other special uses (ammonium to cause the plant to produce flowers and seed; a light application [200 lbs. per acre broadcast] of ammonium sulfate to cause soil to warm up faster in the spring and stay cooler in the summer, if the humus content is adequate).

Foliar feeding. Plants are able to take in many nutrients through their leaves. Spraying materials on the above-ground part of the crop is often called foliar feeding. It is not practical to supply a plant's total needs of the major elements in this way, but foliar feeding is an excellent way to supply secondary or trace elements, to "energize" a crop whose growth is lagging, and to "switch" a crop from vegetative growth to

[8]M. M. Mortland, *Advances in Agronomy*, 1958, vol. X, p. 342–345; R. I. Papendick & J. F. Parr, *Soil Science*, 1966, vol. 102, p. 198; J. F. Parr & F. E. Khasawneh, *Plant and Soil*, 1968, vol. 29, p. 460–470; C. F. Eno, W. G. Blue & J. M. Good, Jr., *Proc. Soil Science Society of America*, 1955, vol. 19, p. 55–58.

[9]*Acres U. S. A.*, May 1981, p. 5.

flower and seed production. And to do so for only pennies per acre of materials in some cases.

Of course you need special spraying equipment (or to hire custom or aerial spraying). The finer the spray the more effective in being absorbed by the plants and the cheaper the cost of materials (less is needed). An ordinary field sprayer does little good. Sprayers that produce a mist or fog are better, and one that homogenizes is best. A highboy tractor allows tall crops such as corn to be sprayed late in the growing season (foliar spraying is often required every 10 days or two weeks).

If foliar feeding is done to supply deficient trace elements, results may or may not be satisfactory. Generally, unless the major and secondary elements are adequate and balanced in the soil, trace element application may do little good. Trace elements come in various chemical forms, both inorganic and organic. Some are better than others, and some are very expensive. However, only very small amounts of trace elements are needed by plants. You should never add trace elements just because you think they *might* be deficient, since too much of some trace elements can be toxic. A soil test should be done and only the recommended amount should be applied. "Shotgun" mixtures of trace elements can be harmful if one or more elements are not deficient. Often trace element deficiencies will magically disappear after you build up soil humus and organisms, since these supply trace elements and break them out of soil minerals.

You can buy pre-mixed foliar sprays, but it is cheaper to mix your own from scratch. The mixing and use of foliar sprays can get very complicated, with special mixes for each crop and situation. We cannot go into that much detail here, but there are some simple but effective foliar sprays that give good results with most crops. When mixing and using foliar sprays, use only clean equipment that has never been used for herbicides and insecticides, and use chlorine-free ingredients and water (or at least no more than 1% chlorine). Soft or deionized water is best; minerals in water may interfere with spray ingredients.

A good beginning spray is liquid fish emulsion, diluted at 2 gallons per 100 gallons water. The fish emulsion may have lumps that would clog a sprayer, so add one gallon of fish to 10 gallons of water, mix and strain through a fine screen or cloth strainer, then finish diluting. Spray this at a rate of 2 gallons per acre.

Another excellent formula is about 3 parts liquid fish to 1 part liquified seaweed (about 6 quarts fish and 2 quarts seaweed per 100 gallons water). Both of these products from the sea are full of trace elements, plus the seaweed has plant hormones and growth stimulants.

Fish also contains about 10% nitrogen. Be certain that these spray mixtures are not alkaline, but between pH 5.0–6.5 (inexpensive pH testing paper can sometimes be purchased from a drugstore). First adding 1 pint to 2 quarts of liquid phosphoric acid to the water will neutralize alkaline ingredients. Spray at the rate of 1 quart per acre of fish-seaweed mixture.

A more complex formula for seed crops in the middle of their growing season is: in 100 gallons water, mix in this order,

1. One to four pints liquid phosphoric acid (40–45%, available from feed stores), depending on level of phosphorus in soil.
2. One gallon household ammonia (substitute 3–4 lbs. calcium nitrate if the crop is still young).
3. If soil is low in phosphorus, add a potassium source, either 1/4 to 3 lbs. of potassium sulfate or 1/2 to 4 lbs. of potassium hydroxide (this is a type of lye; dissolve first in a little water).
4. Five lbs. of colloidal (soft rock) phosphate. Stir into a little water, let settle and use the milky water.
5. Any trace elements that are seriously deficient. Up to 5 lbs. of filtered poultry manure is a good source of trace elements, especially boron. Manganese is a trace element that promotes good seed or fruit development; use if needed (too much is toxic). Most crops can use some iron; too much is not harmful. Leave some scrap iron in a bucket of water and use the rusty water as an iron source.
6. Two gallons of liquid fish emulsion or the fish-seaweed combination covered above is an excellent option to supply trace elements and growth hormones.
7. One quart emulsified oil (crop oil, dormant oil, Stoddard solvent); use no more often than once a month.

Spray the above mixture at a rate of 7 gallons per acre, or enough to wet plants. For non-seed crops (alfalfa, hay, grasses), use 3–4 lbs. of calcium nitrate in place of the ammonia and omit the potassium source. Do not use the trace element manganese. Alfalfa requires high calcium, which is best supplied by liming the soil; adding 3–5 lbs. of finely ground calcium carbonate (high calcium lime) to the spray at step 4 (above) will supply some. Do not add fish emulsion after two weeks before harvest of hay or forages, because cattle do not like the taste.

For root crops and potatoes, increase the amount of phosphoric acid and potassium source. Substitute potassium nitrate for ammonia (from 10 lbs. per 100 gallons water for carrots, to 20 lbs. for beets, up to 50 lbs. for potatoes).

Up to 5 gallons of molasses per 100 gallons water is an excellent addition for sprays (add at step 4, above). It supplies iron, stimulates soil life, and increases palatability and nutrient content of hay crops, especially if sprayed just before harvest.

The figures used above apply to the use of a homogenizing sprayer. For ordinary mist sprayers, the amount sprayed per acre may have to be doubled or more. For crops with large leaf area (soybeans, vegetables) or greater growth, amounts also will have to be increased.

The best times to spray are in foggy weather or in the early morning (4–6 AM), since this is when the plants take in materials best. The evening is another good time to spray, after 7 PM.

A good way to test whether the spray mixture is going to help your crop is to spray several plants with a small hand sprayer. Wait a half hour and test the sugar content of sprayed plants with a refractometer (see later in this chapter), compared to unsprayed plants. If the sugar content increased, the mixture is good.

A spray to "switch" a seed crop from vegetative growth to flowering should contain ammonia or ammonium nitrogen rather than nitrate nitrogen. For small acreage or gardens, a spray of diluted vinegar (2 cups per gallon of water, spray 1 gallon per 100 square feet) should be used about halfway through the growing season. For larger acreage, a light side dressing (100–150 lbs per acre) of superphosphate (0 - 20 - 0) will do the trick. If these aids are used for "switching," be sure the soil has enough fertility to fill out the seed or fruit. Additional side dressing may be necessary. The same thing applies to other growth stimulants, such as seaweed, hormone, and enzyme sprays. It does little good to stimulate growth if the soil cannot supply the needed nutrients.[10]

In review, foliar feeding is not practical for supplying major elements, but has value for supplying trace elements, for pulling a crop through a stress period, and for "energizing" and "switching" in order to hasten maturity.

Soil management. Soil management includes everything you do to the soil: crops grown and their sequence (cropping system), use of fertilizers and pesticides, tillage, irrigation and/or drainage, terracing, etc. Obviously soil management—how you treat the soil—has everything to do with the condition and fertility of your soil 5, 10, and 50 years from now, as well as the yields and quality of your crops, the health of your animals, and your profit/loss figures.

We have already covered a great deal about improving the soil, but now let's say something about cropping systems and tillage.

Cropping systems. There are many different ways of growing crops,

[10]D. Skow & C. Reams, foliar feeding seminars, 1981, 1983; *Acres U. S. A.*, September 1984, p. 24; March 1985, p. 12–14; *The New Farm*, February 1985, p. 32–34.

but they can be boiled down to two basic approaches: growing a sequence of different crops on the same piece of land (rotation) and growing the same crop year after year (monoculture, continuous cropping). Both approaches are championed by some and criticized by others. There are advantages and disadvantages to both; it depends on your particular situation as to which may be best for you. Either one can be unsuccessful if not done properly.

Rotation. The advantages given for crop rotation include its ease of maintaining or building soil organic matter, and supplying much crop nutrition (especially nitrogen) from plowed-under forage crops. Also, rotation helps prevent pest, disease and weed problems by growing different kinds of crops in the same field. Under average soil conditions, rotation can give about 10% higher yields than continuous cropping.[11]

Continuous cropping. The proponents of continuous cropping say that you can match crops with soil type and topography—grow row crops on level land to reduce erosion and keep hilly land in forages or pasture. Or you can specialize and grow the same thing on all your acres, simplifying your machinery roster. Since different crops do best under different fertilizing programs, you can "fine tune" soil to grow top yields of any particular crop.

Disadvantages of monoculture include soil depletion if crops are continuously harvested and if little organic matter is returned to the soil. Row crops and vegetables are especially bad for this. Soil-borne pests and diseases can build up, and weeds may increase. With row crops, erosion can be tremendous. Considerable outside fertilizer may be needed.

Which? Whether you rotate or use continuous cropping depends on your needs, soil, and topography. If your soil is depleted and you have no source of much organic matter to recycle (or compost), rotation is the best way to build up soil, with emphasis on forages and/or pasture grasses. Proper recycling of manure and other organic matter can build up soil under continuous cropping. You can do a great deal by growing and incorporating a cover crop (rye is excellent for this).

Some good crops to add nitrogen to the soil are alfalfa, sweetclover, and red and Ladino clovers (soybeans, although they are legumes, add little nitrogen to the soil because they use most of it in seed production). Grasses and legumes provide a slow release of nitrogen when plowed under. To increase soil phosphorus, grow alfalfa or plow under sweetclover. Buckwheat will increase a number of soil minerals. Perennial grasses are excellent for improving soil structure.

In general, once soil is in good shape, a rotation should be used that

[11]*Crops & Soils*, March 1984, p. 12–14.

grows a nitrogen-increasing crop just before a high nitrogen consuming crop (such as corn), with perhaps a year or two of low nitrogen consuming crops following that (such as small grains or root crops). Then one or more years of soil-building crops should follow (legumes or grasses). In dry climates, avoid moisture-depleting crops such as alfalfa, corn, long-season crops, and shallow-rooted grasses. A fallow season may be necessary for soil to recover from a crop.

After you have built up depleted soil, you may want to consider "fine tuning" some fields to grow top quality forages or high value row crops continuously, with careful monitoring of soil conditions to prevent depletion.

Intercropping. The practice of planting two crops together is not new, and although it has not been popular, interest in intercropping is growing. With a crop such as corn, wide-spaced rows are used and another crop such as soybeans, forages, small grains, or sod is planted between the rows. This has the advantage of excellent weed control, plus the yield of the second crop. If a forage or grass was to follow corn in a rotation, it will already be established. A legume (alfalfa, clover, etc.) will add extra nitrogen in good soil. The yields of corn are not hurt (except that wider rows are used), but the yield of the other crop will be reduced from shading. If soil moisture is low, intercropping would provide too much competition for water, however.

Interplanting a small grain (winter rye or barley) and/or a forage legume (clovers, hairy vetch, etc.) with corn will give good ground cover over winter. In the spring, plowing in the cover crop will provide nitrogen and organic matter for the next crop. The cover crop can be interseeded during the last cultivation of the corn (about a foot high).[12] See also Chapter 7.

Tillage. Tillage is done for three reasons: to change soil structure and prepare a seedbed, to control weeds, and to manage crop residues and organic wastes. Many kinds of tillage methods and machinery have been used over the decades. Some work well in certain types of soil or climate; for example, disk plowing is good for dry climates and light soils, while chisel plowing does not work well for moist soils or sod.

Each type of tillage has its own place and best uses. Moldboard plowing, which has come in for considerable criticism because it inverts the soil and disturbs the soil's ecosystem, still is useful for breaking sod and turning under green manure crops. If the moldboard plow is adjusted so that the slices are stood up on edge rather than being inverted,

[12]*The New Farm*, January 1984, p. 38–40; *Acres U. S. A.*, September 1984, p. 20–21.

most of the undesirable results are prevented. This type of moldboard plowing in the fall, followed by freezing and thawing, is good in heavy soils and can break up a hardpan.

In order to do the most good in improving soil structure and humus content, organic matter must be worked *into* the soil, not left on top. But at the same time, if the lower soil layers are tight and not well aerated, fresh organic matter should not be plowed in that deep, but kept in the upper several inches. And of course in some cases, some trash or a thin mulch should be left on top, such as to prevent wind erosion in the plains region.

In some cases or for some purposes, ridge-tilling is excellent. If organic matter is worked into the upper several inches, decomposition to form humus occurs quickly because the ridges provide better aeration. In northern climates, ridged soil warms up fast in the spring if the ridges run east-west. In dry areas, the ridges catch rain or snow. Ridges are commonly used for irrigation and truck crops. See *The New Farm*, July/August 1984, p. 18–20, and November/December 1984, p. 34–37, for helpful hints.

Spring primary and secondary tillage can reduce the earthworm population up to 75% in the upper eight inches, according to studies at the University of Guelph, Ontario, Canada. Although the worms can restore their population levels by fall, summer is the time when they are most needed. Microorganism populations can also be upset.[13] Shallow spring tillage is best, with whatever deeper tillage is needed being saved for the fall.

To till or not to till? The biggest trend these days is to reduce the number of trips across the field because traffic tends to compact soil and waste time and fuel, and because excess tillage causes soil organic matter to be oxidized faster. Also, the threat of erosion is increased. So the big thing being pushed today by the USDA, university extension, and others is *no-till*, or reduced tillage called by various names: minimum-till, conservation-till, etc. These methods are touted as saving fuel, reducing compaction and erosion, conserving soil moisture, and even building up soil organic matter. Yields are said to be "the same or higher" than with conventional tillage. Scientific studies have proved that this is true. But it is only one side of the story. Listen to the rest.

Leaving the crop residues on the surface, and little or no deep tillage, means that subsoil will soon become depleted in organic matter and become more anaerobic. Some conservation tillage systems incorporate

[13]*Acres U. S. A.*, July 1984, p. 12; October 1984, p. 24.

some of the trash by disking, and deep-till by chisel plowing, both of which are good.

The surface mulch does indeed reduce erosion from heavy rainfall, but it also reduces soil temperature in the spring, delaying seed germination (up to 8 degrees F cooler in Wisconsin).[14] Also it forms a haven for certain insect pests (cutworms, wireworms, corn borers, seed corn maggots, grubs, flea beetles, and billbugs) and plant diseases (eyespot and stalk rot of corn). Weeds tend to go wild.

So what is the recommended remedy for these problems? Zap the pests and weeds with powerful insecticides and herbicides, and use disease-resistant varieties. And because of the high organic matter content of the surface layers, extra high doses of poisons are required! (pesticides are buffered and deactivated by organic matter).

Nitrogen and potassium shortages often occur with no-till[15]; more applied (outside) fertilizer thus may be needed. Toxic substances (phytotoxins) leaching out of some plant residues stunt growth and germination of the next crop. Because of the surface mulch, soil microbial populations are high near the surface, but in deeper soil, beneficial organisms and oxygen levels will be very low (anaerobic conditions). This will lead to many problems, detailed in Chapter 3, including poor drainage, and compacted and toxic subsoil. Crop root growth will be restricted to the upper few inches, increasing the possibility of lodging of crops such as corn.

What about yields? "In northern climates on poorly drained, clay soils, the no-till system will likely yield significantly lower than conventional tillage."[16] In well drained soils, yields are similar to those with conventional tillage, and only on unirrigated, droughty (sandy) soils or in southern climates does no-till outyield conventional methods.[17] No-till or plow-plant systems may be able to be used effectively in such cases as double cropping, with a crop such as soybeans following a small grain, or where one row crop follows another and a surface mulch is needed to prevent erosion.

So why, with all these disadvantages and problems, are no-till and other reduced tillage systems being promoted so much? Could the higher amounts of pesticides and fertilizers have anything to do with it? The universities *do* receive grants from the chemical companies to do research. Hmmm, I wonder. . .

Alternatives. Are reduced tillage systems worth it? Do the dollars

[14]University of Wisconsin Extension publication A3091, 1980, p. 10.
[15]*Ibid.*, p. 11–12.
[16]*Ibid.*, p. 13.
[17]*Ibid.*

saved on fuel outweigh the long-term soil damage from increased toxic poisons and anaerobic subsoil?

Sad to say, the desirable goals of reduced erosion, moisture conservation, weed and pest control, and high yields can be reached by following the soil improvement methods given earlier in this chapter, without poisoning the environment with pesticides. Humus and a healthy population of soil organisms will provide proper plant nutrition and help control pests and diseases. Weeds will not be a problem on balanced, well-aerated soil. And you can even save fuel, because soil with good tilth is easier to work and doesn't compact easily. Leaving some "trash" (crop residues) on the surface is excellent for reducing erosion.

You don't have to be a slave to the system.

Are hybrids better? One of the triumphs of modern agriculture is the development of hybrid varieties of crop plant species. Hybrids have given us greatly increased yields, helping to alleviate world hunger. The "green revolution" was built on recently developed hybrid varieties of wheat, corn, and rice. Plant breeders can "tailor make" almost any type of hybrid they want, and some varieties of garden vegetables are truly amazing. But are the hybrid marvels of modern technology really better than the original open pollinated varieties?

A hybrid results when parents from much different ancestry are crossed. This brings the chromosomes and genes (which control the plant's characteristics) from the two parents together into a new combination. Thus the hybrid has whichever set of characteristics the plant breeders choose. They can design a hybrid that is drought resistant, or one that has a higher starch content in the grain, or one that matures earlier. They find a plant or variety that has the desired characteristic and then grow an "inbred line," a variety that "breeds true" for that characteristic. Then they cross that variety with a different inbred line that has some other desired characteristic or characteristics. A "single cross" hybrid is the result. Sometimes two different single cross varieties are crossed to give a "double cross." Inbred lines are generally inferior in certain ways, but when they are crossed with another inbred line, the resulting hybrid has "hybrid vigor," a dramatic increase in yield and other desirable characteristics.

However, when seed from a hybrid is planted, not all of the next generation has the desired combination of characteristics, but a certain percent of individuals revert back to one of the inbred parents' characteristics. So to grow hybrid crops you must buy new seed every year. In contrast, the "old fashioned" open pollinated varieties "breed true," and you can save your own seed from the best plants and gradually develop your own variety, adapted to your own soil and climate.

Hybrids are engineered to be uniform and to produce excellent yields under a certain limited set of conditions, such as low calcium soil in a cool, wet northern climate. But in other conditions, such as a hot, dry year, the crop is a disaster. Open pollinated varieties contain more variability in their genes and can grow well in a variety of weather and soil conditions.

The overwhelming goal of the plant breeders is high yields—that is, quantity, not quality. Each hybrid variety is tailor made for a certain kind of sick, out-of-balance soil. Few hybrids are bred to produce truly high quality food and grow on good healthy fertile soil. Since hybrids were first introduced in the 1930's, the protein content of crops such as corn and wheat has been steadily declining. This also reflects the depletion of our soils. When hybrid grain is tested for nutrients, it may lack certain trace minerals or have a poor balance of amino acids. Animal raisers often note that their animals grow faster and are healthier when fed feed made from open pollinated crops.

A recent "genetic triumph" is triticale, a hybrid between wheat and rye. It is boasted to have the high yields of wheat and the hardiness and disease resistance of rye, and to be very nutritional. But because it is a hybrid between two species, the cell nuclei of the grain are abnormal, and the seeds are usually shriveled because of abnormal development of the food storing tissue (the endosperm).[18]

A good open pollinated variety, when grown on good, fertile soil, will often yield about as well as a hybrid, but the big difference is QUALITY. The open pollinated varieties win hands down. Many farmers are abandoning hybrids and switching to open pollinated crops, especially if they are feeding them to their animals. If your soil is in good shape, you may be in for a pleasant surprise if you try an open pollinated variety.

Quality. Does the quality of a crop really make a difference? Isn't a high-yielding variety better than a lower yielding one? In Chapter 2, we showed that animals do much better on food grown with "natural" farming methods. If you grow crops to feed your animals or for human food, you probably appreciate that quality is important.

But if you are growing cash crops, the market system is set up to pay for quantity—bushels or pounds. It shouldn't be that way, and there are some specialty markets that will pay a premium for quality, such as foreign buyers, poultry feed, zoos, some bakeries, and the health food market. But even when selling by bushels, quality still does make a difference, because high quality grain has a higher test weight, and the bushels are measured by weight, not volume. So, then, striving to grow high

[18]*Science*, 1982, vol. 216, p. 1307.

quality crops is a worthwhile goal in any case. Even if you or your animals don't eat your crops, someone else or their animals will, and a conscientious food producer should do the best job he can.

How *you* can measure quality. There are various lab tests that are supposed to measure quality, such as crude fiber, acid detergent fiber, digestible protein, and crude protein. They all measure just one particular quantity which is comparatively easy to measure, whether it is fiber or protein or what, and use that measurement as an *index* or approximation of food value or quality. The commonly used test, crude protein, can be very misleading, since it doesn't even measure protein, but only measures nitrogen content of the sample. This figure is then multiplied by 6.25, assuming that proteins contain 16% nitrogen (16% = 6.25). But if the crop was grown in unbalanced soil or under stress conditions, its metabolism may have been upset, with other nitrogen-containing chemicals accumulating in its tissues, such as nitrates, ammonium, amides, and amino acids. These are not protein, and they would give a falsely high crude protein reading. Nitrates are quite toxic in too great amounts.

A simple index such as the above measurements, cannot measure true nutritional value of a crop. There are just too many nutrients involved: carbohydrates, fats and oils, proteins, vitamins, and minerals. Those are the main types of nutrients, but there are hundreds of different specific kinds of proteins, for example, dozens of carbohydrates and fats, over a dozen vitamins, and almost as many minerals, plus other poorly understood "growth factors." Not only their presence or absence is important, but their *balance* is extremely important. This is well known in the case of proteins, in which the animal's diet needs a certain balance of amino acids, which are the building blocks of proteins. Also, certain vitamins and minerals interact in the body. For example, in humans, vitamin D is most effective when there is also adequate vitamin A, choline, vitamin C, calcium, and phosphorus in the body. The trace mineral selenium works best with vitamin E.

A better way to measure food quality is to feed it to animals and see how they do—growth, health, reproduction, and meat, milk, or egg production. This is how the above-mentioned laboratory measurements were arrived at. But such tests are expensive and time-consuming, and you may not want to risk losing any of your valuable livestock. There are simpler and quicker ways you can measure crop quality.

In general, the higher the *test weight* of crops, the better the quality. Low quality, watery grain or hay will dry down to almost nothing, while plant tissue that is packed with protein and minerals will have less water. The old "standard" bushel of corn weighs 56 pounds, but these days there is a lot of 52, 50, or 48 pound corn being grown. No wonder livestock and poultry are easy prey to diseases and have to be fed rations fortified with the vitamins and minerals that should be in the crop.

One of the handiest tools to measure quality and monitor crop growth is a refractometer. This is a precision instrument which can quickly measure the percent sugars in a plant's sap, while you are standing there in the field. In most cases, the higher the sugar content, the higher the protein and mineral content, and the food value. Relatively inexpensive, refractometers are routinely used in the food industry, by canneries, wineries, and breweries for example, to measure the quality of the fruits and vegetables they buy from the farmer or of the foods and drinks they manufacture.

Using a refractometer is easy. Just squeeze a few drops of juice from the stem, leaves, or fruit of the plant onto the glass prism of the refractometer, close the "lid," and look through the eyepiece. The sugar content is read on a numbered scale in units called Brix (same as percent).

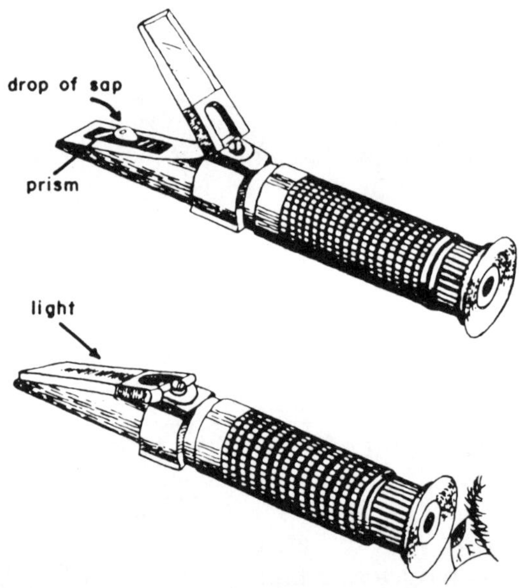

Using a refractometer is easy. A drop or two of sap from the plant is squeezed onto the glass prism. Then the "lid" is closed; the instrument is held in the light, and the sugar content is read on a scale visible through the eyepiece.

Brix (sugar) Readings for Some Plants

	poor	*average*	*good*	*excellent*
alfalfa	4	8	16	22
carrots	4	6	12	18
grains	6	10	14	18
corn stalks	4	8	14	20
corn, young plant	6	10	18	24
grapefruit	6	10	14	18
lettuce	4	6	8	10
sweet corn	6	10	18	24
watermelon	8	12	14	16

By comparing with standard levels (see table) or with past readings you have made, you can see how your crops measure up that day. There are a few things to be careful of, however. The sugar content will vary from one plant to another, from one part of the plant to another (higher in leaves and upper stalk), in different weather and times of day (higher on warm, sunny days in the afternoon), and for sick and healthy plants (sometimes a sick plant will give a high reading if the sugar isn't being carried down from the leaves into the fruit or grain as fast as it should; a healthy plant should have a low sugar reading in the early morning, while a sick one may be high). So you should check the sugar reading at the same place in the plant (say middle or upper leaves, or middle stalk) and at the same time of day on sunny days.

Refractometer readings may not correlate with the standard lab tests, such as protein tests, but unless the crop is obviously sick, the higher the refractometer reading, the better the true food value.

Refractometers can also be used to test animal or human urine and to test the quality of garden or grocery fruits and vegetables.

CHAPTER 6

The Future— The Scientists' Vision

"Ever learning and never able to come to the knowledge of the truth." II Timothy 3:7

Farmer Brown settles down in his office to do his day's work. He surveys his farm through the picture windows of his air-conditioned office. He farms 5000 acres, with two specialties: (1) dairy and the hay and grain crops the cattle consume, and (2) algae, a protein source for human consumption. His farm is small compared to his neighbor's, who grows vegetables for the nearby sprawling city, a megalopolis with 50 million inhabitants. The year is 2085.

First, Mr. Brown checks the conditions of his hay and corn fields. He speaks to his desk-top microcomputer and tells it to read the data from

the remote sensors implanted in different parts of his fields. Instantly rows of figures flash on the giant display screen, giving the soil moisture, pH, nutrient content, crop sugar content, and several other factors. Another oral command makes the computer show a satellite infrared picture of his fields, allowing him to spot areas of stressed or diseased plants. A read-out of the planned weather for the week is also checked.

Mr. Brown plans the field work for the day, aided by the data banks stored in the computer's memory concerning soil conditions, nutrient levels and crop needs, soil types, fertilizers applied, and hybrid crop variety planted.

The corn fields had been planted with seed treated with insecticides and fungicides to control pests and diseases that attack seeds and seedlings, and an insecticide that is carried up into the growing plant to kill pests of the older plant. The special hybrid variety is resistant to several diseases, produces very high amounts of protein, and develops nitrogen-fixing nodules on its roots like legumes. It is a perennial variety of corn, meaning that it grows up each year from permanent rootstocks; thus no yearly planting and tilling activities are required. Remote-controlled driverless tractors were used to till-plant the fields, following wires buried in the ground. All machinery that travels on the fields drives on the same paths to reduce soil compaction on the parts of the fields used for crops. Built-in computers and sensors operate the tractors at maximum fuel efficiency. Large till-plant rigs covered 24 rows at a time.

The corn and hay are harvested by huge combines which automatically follow the cut edge and adjust the cutting height using sensors on the gathering platform. Corn is picked, shelled, weighed, and stored in

bins on the combine, while stalks are chopped and disked into the soil by a trailing tillage tool bar. Hay is cut, processed into cubes ready for the dairy cattle, and stored in the combine's bins.

Mr. Brown's dairy cattle have been bred to produce large amounts of high butterfat milk. They were produced by using hormones to cause their mother to ovulate dozens of eggs. She was then artificially inseminated with semen treated to contain only female-producing (X chromosome) sperm. The dozens of embryos that grew were then removed from the uterus and split into hundreds of embryos. These were then transplanted into receiver cows which carried them to birth. In this way, many genetically identical cattle superior in milk production were produced.

Small electronic transponders implanted under the skin identify each cow with a number. Implanted sensors detect the animal's body temperature and reproductive system status. When the cow is hungry, it approaches the computerized feed dispenser. The transponder activates a feed-dispensing sensor; the main computer checks the animal's milk production record and reproductive status, then tells the dispenser to dump out the exact amount of the right feed ration. Every time the cow is milked by the totally automatic milking machines, the milk is automatically weighed and tested. These data and the cow's number are recorded by the farm's computer and a complete record is kept of production.

Milk is processed into cheese, buttermilk, powdered milk, irradiated non-refrigerated milk, and whey by-products by a sophisticated factory on the farm. The cattle are housed in multi-level "high rise" barns, with a feed storage facility on the roof.

Manure from the barns is automatically washed into tanks, where methane is first produced from it. Then the solid residues, containing fiber and nutrients, are dried and pelleted for addition to the dairy feed rations. Waste heat, carbon dioxide, and manure liquids from the barns are piped into Mr. Brown's dome-like algae-growing greenhouses, where they warm and nourish the billions of microscopic algae cells floating in large water-filled tanks. Carbon dioxide and special nutrient solutions are bubbled through the tanks. The sun in the day and lights at night stimulate the algae to carry on photosynthesis 24 hours a day. Oxygen given off by the algae is piped into the dairy barns. Automatic flushing and straining equipment "harvests" the algae; then it is dried, packaged, and transported to the city. Algae is a nutritious and cheap dietary supplement, and it comes in several flavors and in the form of flour.

Mr. Brown's neighbor, Mr. Smith, runs a 12,000 acre vegetable farm. Gigantic "robot farmers" hundreds of feet long do all of the work.

They consist of narrow frameworks mounted on wheels. They slowly move down concrete tracks in the fields, guided by lasers. Depending on needs, machinery attachments are computer programmed to till-plant, spray fertilizer or pesticides, or harvest. These attachments move back and forth along the framework, covering 12 rows at a time. Irrigation water is supplied by a type of trickle irrigation, with plastic tubing distributing precious water to underground "nozzles" called emitters. Daily weather data on wind, sunlight, temperature, and humidity are used by a computer to determine how much water the plants lost through their leaves, and thus how much to release through the irrigation system. Much less water is used than with the old-fashioned ditch or sprinkler systems. During harvesting, sensors detect the ripe vegetables and robot-like fingers pick them; conveyers carry them to packaging units. Special hybrid varieties are engineered to grow and mature in half the normal time, to be uniform in size, and to mature all at the same time. Buried electrical heating wires allow vegetables to be grown through the mild winter.

For high-priced tropical fruits and vegetables, Mr. Smith uses large slowly rotating drums in huge buildings. The plants grow out of slits in the drums and roots are periodically sprayed with nutrient solutions. Lights allow for 24-hour growth. Multiple stories of the tanks allow large amounts of food to be grown on small acreages.

It's already here! Does all of this futuristic description seem totally fantastic, like science fiction, like wild imagination? Do you say such developments will never come?

I have news for you. All of the things I described are either being used now, are being researched, or are in the planning stages. This is the kind of agriculture the scientists and technologists have planned for us in the next 50 to 100 years.

Is this the kind of world you or your children want to live in? Super-large farms. Super efficiency. Every possible aspect of growing crops and animals controlled by computer and automation.

Is this the way it will turn out? Before we comment on that in the next chapter, let's look in more detail at a few of the developing technologies that the researchers are working on and the trends of today's society which will influence the future.

Larger farms. For over 100 years there has been an inexorable increase in size of the average American farm, and at the same time, a reduction in the number of farms and percent of farmers in the population. The number of acres under cultivation has remained nearly the same since at least 1920. An average of 2000 farms have gone out of business *every week* since 1950. In 1920, farmers made up 30% of the population; in 1950 they were only 16.6%, and in 1980, a mere 3%. Less than one-fifth of families living in rural areas make their living from farming. Farm families as a whole only derived a little over one-third of their income from farming in 1980; the rest came from off-the-farm jobs. In fact, small farmers (annual farm sales of less than $20,000) earned 88% of their income off the farm. Only 4% of U. S. farms are large (annual sales of over $200,000), but they accounted for half of total cash receipts and 20% of net farm income in 1980. Over 20% of farmers' incomes were below the poverty level. Although there has been a "reverse migration" of urban dwellers into rural areas since 1970, with a slight increase

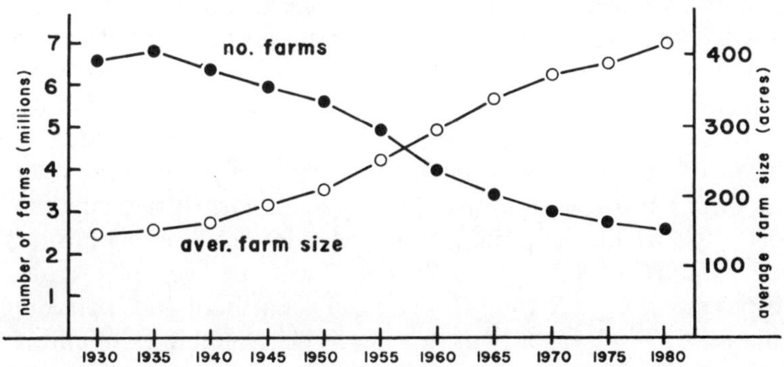

The trend over the last 50 years has been for the number of farms to drop drastically, but for the size of farms to increase.

in small farms (under 50 acres), still the trend of fewer farmers and larger farms is expected to continue in the future.[1]

There are several factors that have contributed to these changes in American agriculture. The pioneering settlers, toiling with their hands and animal power, could barely feed their own family. Since Eli Whitney's cotton gin in 1793, Cyrus McCormick's grain reaper in 1834, and so on, mechanization and technological advances have increased the farmer's muscle until today the American farmer produces enough food to feed himself and 78 others. In 1950, the figure was only 16 other people. The early farmer was pretty much self-sufficient, raising his own food, recycling manure back to the fields, and buying little in town. After World War II, the rapidly increasing use of hybrid crop varieties, insecticides, and commercial fertilizers meant that a considerable amount of profits had to be spent for outside "inputs." The figure is now about 70%.

Beginning with the New Deal programs in 1933, the government and policy makers have exerted an increasing control over farmers. Laws and regulations control what crops are planted, soil conservation practices, and manure disposal, for example. Market regulations influence the price the farmer receives for his products. Since 1950, farm commodity prices have been held below "parity," the level at which farm income would be comparable with the other segments of the economy. Since then, farm debt has soared (from 10% to about 20% of equity), spurred on by inflating costs of fuel, machinery, fertilizer, and just about all the other "inputs" considered essential to modern farming. Just between 1972 and 1982, fertilizer and lime costs increased 333%, seed costs increased 350%, and interest payments soared 376%![2]

Many farmers have "gone under" and most of the rest are slaves to the banks and credit organizations for operating loans. High interest rates will help insure that 4 or 5% of the nation's farmers will continue to go out of business every year. From 1977 through 1982, prices paid by farmers for various goods, services, interest, etc., have increased an average of 10% per year, while prices received for farm products rose by less than 7% annually. Any businessman will tell you it's hard to make a profit under those conditions! Government farm price support programs, supposed to help the farmer, have favored the big farmers and hurt the small (in 1983, only 4.6% of the farms received 25% of all direct farm payments), increasing the gap between them and increasing the likelihood of small farm bankruptcies. The available farmland is then

[1]*Agriculture 2000, A Look at the Future*, 1983, p. 91.
[2]*Farm Paper Letter* (USDA), February 17, 1984.

gobbled up by the big farmers. The bankers and politicians frankly admit that the severe economic conditions will drive out the "less efficient" farmers.

The amazing increase in productivity of modern farming methods is often blamed for the farmer "producing himself out of business." Surpluses of grains, dairy products, and other commodities are the reason given for low market prices and set-aside programs. Is the farmer to blame for his woes, or is he a victim?

A growing number of farmers, farm organizations, and even a few politicians, economists, and bankers are beginning to believe the latter. Could government and economic policies be designed to drive most farmers off the land so the land can be reorganized into large, efficient food factories? In the fall of 1983, U. S. Secretary of Agriculture John Block told the *Des Moines Register* that according to USDA plans, there must be a further 16% reduction in the number of farmers, which means nearly 400,000 more farmers will go out of business.[3] More details are given in Chapter 8 of *The Rest of the Story. . . about Agriculture Today*.

Machinery. Going along with large farms is large and sophisticated equipment to till, plant, and harvest so much land in a short time. More efficient diesel engines are being developed which will use some of the exhaust heat to run a turbine to pressurize intake air to give extra power or save fuel. Future engines may have high temperature parts made of ceramics, and with special lubricating oils, the entire cooling system (fan and water pump) can be eliminated. Computerized sensors and automatic engine controls to regulate the air/fuel mixture can operate a tractor for maximum fuel efficiency, or for maximum power, or for fastest rate of work, whichever is desired. Similar systems are already found on some automobiles.

Automatic guidance systems, to eliminate the driver's boring job, have been studied. For precision row-following, buried wires which are detected by sensors are best. A less precise system is to guide a tractor by radio transmitters located in three corners of a field, aided by a computer. Sensors that can follow a previously cut furrow or the cut edge of a standing crop are also being developed.[4]

Sophisticated equipment that combines seedbed preparation, planting, and fertilizer and/or pesticide application into one operation is already in common use. In the future, such machinery is likely to get larger and more effective.

The USDA and several universities are developing an extra wide-

[3]*The NFO Reporter*, October 1983, p. 12.
[4]*Agriculture 2000, A Look at the Future*, 1983, p. 116–132.

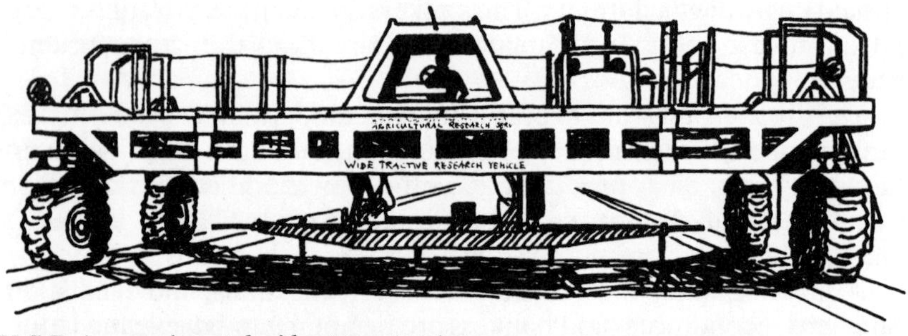

U.S.D.A. experimental wide-stance tractor.

stance tractor (33 feet wide) that will drive only on certain paths through fields, in order to eliminate soil compaction. In experiments, yields have been increased 20 to 100% in uncompacted soil.[5] This type of approach leads to the "robot farmer" described earlier in this chapter, which has been proposed by Batelle Pacific Northwest Laboratories.

More sophisticated combines and harvesters are certain to emerge from today's technology. The trend will be toward uniting several operations, such as harvesting soybeans, extracting the oil, and processing the meal into seed cake; or processing alfalfa into pellets or cubes; or picking, cleaning, and packaging fruits or vegetables. Automatic sensors would recognize the rows, or cut edges of grain, or tell a mature tomato from an immature one.

Better crops. In the scientific quest to fight world hunger and boost the farmer's productivity, increasing crop yields is the primary concern. There are several avenues of research attacking this problem which are expected to lead to agricultural triumphs in the coming decades. First, there is the manipulation of living organisms, cells, and genes, collectively called biotechnology.

Biotechnology. The name biotechnology includes several techniques now in the forefront of agricultural research. They seek to change living organisms to aid man to grow more food. More than 175 new research and development companies, plus most of the old established agribusiness corporations (including Allied Chemical, Du Pont, Dow, Exxon, Monsanto, Atlantic Richfield, Shell, Ciba Geigy, Upjohn, etc.) are spending millions of dollars.[6] Some specialized techniques of manipulating cells and genes are being patented. Gone are the days when an agricultural scientist merely works with test plots and different combinations of fertilizers.

[5]*The Grower*, October 1982, p. 28.
[6]*Farm Chemicals*, September 1982, p. 33.

Biotechnology involves various methods of working with genes and cells in order to control or change the characteristics of microbes, plants, or animals and to produce useful products using living organisms. These include (1) protoplast regeneration and cloning, (2) somatic hybridization, and (3) gene splicing, often called genetic engineering. Although the names may be foreboding, you can understand something about them.

Protoplast regeneration. By removing small chunks of tissue from a plant or animal's body, separating the individual cells, removing the cell walls, and putting the "naked" cells (protoplasts) in a special growth medium (an artificial "soup" of nutrients), the cells will divide and grow into new clumps of "tissue" (not always normal tissue at this point). This is called *tissue culture*. The cultured cells can be exposed to various chemicals, perhaps to identify cells that are resistant to a certain herbicide or to a virus disease, or to identify cells that can grow with (or without) a certain nutrient or that can tolerate high temperatures or salty water. Desirable cells can then be taken and cultured on a new growth medium. Some will begin growing into roots and leaves, a tiny plantlet. This process of producing new plants (or animals) from cultured cells is called *cloning*. If new individuals are regenerated from the same culture of cells, many genetically identical plants or animals can be produced quickly and cheaply, without going through the usual process of sexual reproduction.

Researchers are now working to develop dwarf varieties of apple trees (which do not have to be grown by grafting onto other rootstocks). Dwarf fruit trees can be picked and sprayed more easily than large trees, and more can be planted per acre. Cloned trees could be produced at a cost of less than one dollar each, compared to $4–7 for grafted trees.[7]

Somatic hybridization. Taking these methods of working with cells one step further, it is possible to take two "naked" cells (protoplasts) from two different species of plant or animal and fuse them together, producing a hybrid having the genes of both parent species. A potato/tomato hybrid has already been produced in the lab. If it were developed for commercial use, the "pomato" could bear tomatoes above ground and potatoes beneath! Under natural conditions, the two species would not be able to cross, but the techniques of somatic hybridization overcome the natural barrier between species. Once a desirable interspecies hybrid is made, it could be propagated by cloning.

The potentials of somatic hybridization fire the imagination of scientists—and science fiction writers. What would a cat-mouse hybrid

[7]*The Grower*, August 1982, p. 18–20.

produce? Or a chimpanzee-human cross? There has been considerable debate on the ethical and moral aspects of such research.

Genetic engineering. The characteristics of all living species are controlled by the genes found in their cells—everything from leaf size to protein content of grain to drought-hardiness is influenced by genes. There are thousands of different genes that determine the characteristics of any one species. They often cause your children to have the same hair color, or musical ability, or tendency to develop diabetes as you do. Wouldn't it be nice to be able to get rid of a gene for diabetes or feeble-mindedness, or to add a gene for athletic ability or genius-level intelligence? That is the basic approach of genetic engineering research, also called recombinant DNA or gene splicing. The previously discussed methods of cloning and somatic hybridization simply use what genes are already present in the cells, but genetic engineering aims to change the genetic makeup of species to suit man's desires.

With present techniques, this is done by transferring an existing gene or genes from one species into another, but future developments may allow genes to be "zapped" by a laser beam or to be "built" from scratch.

So far, microorganisms have been used to transfer genes, either viruses or a type of parasitic bacteria called Ti-plasmids. These microbes normally invade the cells of plants and animals and inject bits of genetic material, so it is comparatively "easy" (using complex modern lab techniques, that is) to insert one or more genes from another species into the virus or Ti-plasmid and let them transfer the gene(s) into the desired organism. Once the gene(s) have been successfully transferred, or recombined (thus the name recombinant DNA) within the receiver species, the standard techniques of tissue culture and cloning can be used to produce many new genetically engineered plants, animals, or microbes.[8]

Genetic engineering is still in its infancy (since the 1970's), but its accomplishments already are amazing. Human insulin-producing genes have been transferred into bacteria, allowing the bacteria to be cultured by the billions and to produce large quantities of human insulin cheaply, thus benefiting diabetes sufferers. Synthetic human growth hormone is helping dwarfs grow to normal height. Similar methods have been used to produce a vaccine for livestock foot-and-mouth disease and to produce interferon, a substance that fights virus infections. The "biosynthetic" interferon has already been tested on cancer patients. Genes in cattle that control production of a hormone that increases milk produc-

[8]*Science*, 1982, vol. 216, p. 1305; 1983, vol. 222, p. 815–821.

tion have been put into bacteria, making possible a feed supplement that boosts milk production.

Developments that are just around the corner include producing large quantities of rennin, an enzyme used in making cheese, casein, and other products, which is currently obtained from the stomachs of calves. Other probable biosynthetic products are vaccines for blue tongue and hog diarrhea, inexpensive production of antibiotics, and pesticides that are from 10 to 100 times more potent than present-day types. Already genes for rat and human growth hormones have been transferred into mice, resulting in king-size "super mice." Now researchers are working to do the same in cattle and sheep in order to obtain large meat animals.[9]

Research is just beginning on engineering crops that are stress-resistant (such as to drought or cold temperatures), disease-resistant, herbicide-resistant, that can grow in salty irrigation water (a serious problem in the western U. S. and many other parts of the world), that can utilize fertilizer nutrients more efficiently, and that produce bigger fruit or more nutritious grain. USDA researchers have discovered a way to control a gene in soybeans that can increase the amount of the amino acid methionine, increasing the nutritional value considerably.[10] Two of the biggest areas of research are (1) improving photosynthesis efficiency, and (2) nitrogen-fixation.

Photosynthesis enhancement. In the miracle of food production called photosynthesis, green plants capture the energy of sunlight and transform it, through a series of chemical steps inside their cells, into sugars, which can later be changed into other kinds of food, such as starches, proteins, and fats. The raw materials of photosynthesis are carbon dioxide from the air, and water. These two simple materials, chemically abbreviated CO_2 and H_2O, are combined to form $C_6H_{12}O_6$, which is chemical shorthand for the sugar glucose. Light provides the energy, and special enzymes and energy-carrier molecules within the cell are the "spark plugs" (catalysts) that help the process along. Enzymes and energy-carriers need certain trace elements or major soil nutrients in order to function.

The rate at which photosynthesis occurs (its efficiency) is determined by several limiting factors: the supply of carbon dioxide and water, the light intensity, the amount of chlorophyll (the green pigment that traps light), and the temperature. Most land plants are less than 15% efficient at peak production. The more efficient crop plants only use about 2% of the available sunlight during the growing season. If only

[9] *Ibid.*, 1982, vol. 218, p. 1298.
[10] *Maryland Farmer*, December 1984, p. 10.

a small increase in photosynthesis efficiency could be made, tremendous progress could be made in alleviating world hunger.

Some algae are as efficient as 20%. Scientists have been busy for years in studying every detail of the complex process of photosynthesis, in algae and in many other plants. They have found that some plants perform photosynthesis differently from others. Most plants, called C-3 plants, first change carbon dioxide into the 3-carbon substance PGA (phosphoglyceric acid) before changing PGA into sugar. A few plants, including corn, sorghum, and sugar cane, change carbon dioxide into 4-carbon malic and aspartic acids instead. These C-4 plants are more efficient than the C-3 species. Perhaps research will find a way to improve inefficient crops. Even certain varieties of a crop species are more efficient at photosynthesis than others; these may be used for commercial hybrids rather than less efficient ones. Breeding plants to have a greater proportional leaf area, faster early growth to put out light-capturing leaves over the ground's surface, more erect leaves to capture light more efficiently, and delayed aging and dying are other approaches being tried.[11]

Some simple techniques are already being used in greenhouse culture. By keeping plants at optimum conditions of temperature and humidity, by increasing light, and by releasing extra carbon dioxide (the air only contains about 0.03% CO_2), remarkable increases in yield have been attained. Tests of releasing carbon dioxide over alfalfa fields have noticeably increased yields, but not as much with wheat, rice, and barley.[12]

Nitrogen-fixation. Most agriculturalists know that those plants called legumes (such as peas, beans, alfalfa, and clover) have small nodules on their roots that take (fix) nitrogen from the air and make it available to the plants. At least, legumes are supposed to have root nodules. Few farmers have ever dug up some soybean or alfalfa roots to see if *their* crops have them or not. Quite possibly the nodules aren't there, or if they are present, they aren't healthy (a pink color inside indicates a healthy nodule). This is because soil conditions are wrong for the special nitrogen-fixing bacteria inside the nodules to survive; perhaps anaerobic or toxic soil.

Anyway, if healthy nodules are present, the crop and the soil receive lots of free fertilizer. Wouldn't it be great if other crops than legumes could grow nitrogen-fixing nodules on their roots too? Then farmers could save millions of dollars on nitrogen fertilizers. This is the goal of considerable agricultural research.

[11] J. W. Rosenblum (ed.), *Agriculture in the Twenty-first Century*, 1983, p. 363.
[12] *Science*, 1977, vol. 196, p. 640; 1984, vol. 225, p. 801–808.

Nitrogen-fixation can be done by several types of organisms. The one found in legume nodules is a bacteria called *Rhizobium*. Another bacteria has been found in root nodules of alder trees, and recently a bacteria called *Spirillum lipoferum* has been found in and near the roots of tropical grasses and corn, although it does not live in nodules. A type of blue-green algae lives inside the leaves of a small aquatic fern which has been used for centuries to fertilize rice paddies in Asia. Root nodules formed by actinomycetes have been discovered in eight plant families.[13] These are symbiotic nitrogen-fixers; that is, they live with another plant. There are also free-living or non-symbiotic nitrogen-fixers which live in healthy soil apart from any plant. These include the bacteria *Azotobacter*, *Klebsiella*, and *Clostridium*, plus several types of blue-green algae.

Scientists have already located the genes that fix nitrogen in bacteria (called the *nif* genes, for nitrogen-fixing) and have transferred them into non-nitrogen-fixing bacteria. They hope to someday be able to transfer the nitrogen-fixing capability directly into the cells of crop plants and tremendously boost world food production. Other avenues of research include getting root nodule bacteria to produce nodules on non-legume crops, especially corn, which is a heavy nitrogen user; and finding strains of nitrogen-fixing bacteria that are more efficient or that fix nitrogen continuously rather than occasionally.

Some progress is being made along these lines. Corn roots have been successfully inoculated with the tropical nitrogen-fixing bacteria *Spirillum*, but *Spirillum* fixes nitrogen best at warm tropical temperatures. Corn in temperate climates may not benefit from it.

Growth regulators. Another area of research that scientists hope will improve food production is growth regulators. These are certain chemicals, some of which are man-made and some of which are produced naturally by plants, which affect the rate of growth and maturation of plants or fruit.

Plants normally produce small quantities of substances called hormones or growth regulators. They either stimulate or inhibit plant growth. For example, auxin is produced in the tips of growing stems; it promotes lengthening of the young stem and inhibits growth of side buds, among other things. Another is a gas, ethylene, which promotes the ripening of fruits.

Recently, a natural regulator was discovered in a member of the mustard family and named brassinolide. Another one named triacontanol was originally found in alfalfa, but occurs in many other plants. They are growth promotors. Two man-made chemicals have been pro-

[13]*Ibid.*, 1975, vol. 189, p. 368; 1977, vol. 196, p. 641; 1980, vol. 209, p. 279–281.

duced which act the same as natural brassinolide. In tests, these growth regulators have increased the yield of radishes and lettuce 15 to 30%, tomatoes 65 to 72%, and beans up to 90%. When used on sweet corn seedlings, there were 20% more ears, 6 to 19% longer ears (with more kernels), and a 51 to 54% increase in fresh weight. A 15 to 20% yield increase of field corn resulted.[14] Two commercial growth regulators are being used to cause young apple trees to bloom one or two years earlier than usual. Also, the apples have better color, ripen uniformly, and do not drop off the tree early.[15]

Planting. Some revolutionary developments are being made in ways of planting certain crops. A new twin-row corn planter plants two close rows (7 1/2 inches apart) spaced 30 inches from the next two rows. Within each of the two rows, seeds are placed very precisely 14 inches apart in an alternating zig-zag pattern, giving the seeds the maximum distance from each other (about 10 inches on the diagonal). This allows very high plant populations (over 30,000) and higher yields per acre. If a 30,000 population were planted in single rows, the seeds would be about 7 inches apart.[16]

It has been found that by treating moistened soybean seeds in polyethylene glycol, germination of old or damaged seed is improved. It will probably do the same for peas, lima beans, and other crops.[17] Soaking tomato seeds 24 hours in an aerated potassium phosphate solution starts the germination process. Then the "primed" seeds are dried and sown by conventional planters; they then germinate much more rapidly than ordinary seeds.[18]

A British company has developed a new 10-row semi-automatic planter for transplanting vegetable seedlings in blocks of soil (especially celery). The machine carries racks of trays holding 46,000 seedlings, with human operators feeding them into the 10 planting mechanisms, which precisely plant each seedling. Over 200,000 seedlings can be planted in six hours, double the amount of standard planters.[19] A similar semi-automatic planting system being used in California transplants vegetable seedlings in small plugs of peat moss about 1/2 inch wide by 1 or 2 inches long. The seedlings are grown by the thousands in trays that hold 300 or 400 plugs. An operator loads the trays on an automatic handling mechanism that dispenses the plugs to the planter.[20]

Another new development is called fluid drilling. Newly germinated

[14]*Ibid.*, 1981, vol. 212, p. 33–34.
[15]*The Grower*, July 1982, p. 14, 16.
[16]*Farm Industry News*, July/August 1982, p. 16, 18.
[17]*American Vegetable Grower*, September 1981, p. 66.
[18]*Ibid.*, December 1981, p. 6–7.
[19]*Ibib.*, February 1982, p. 8, 10.
[20]*Ibid.*, May 1982, p. 8–10, 12, 14.

seeds of vegetables such as celery, tomato, carrot, pepper, and asparagus, are suspended in a gel, then precision planted in a furrow or through a plastic mulch (a layer of black plastic covering the row). Plants emerge earlier, grow faster, and yield higher fresh weights (35–119% for tomatoes, 50% for celery). Experiments to add fungicides or competitive beneficial fungus spores to the gel to control damping off or other diseases may allow reduction in field operations and savings in time and money.[21]

Some vegetable seeds, such as tomato, are being planted by machines that bury a narrow plastic tape with the seeds affixed on it at precise spacing. The tape dissolves from soil moisture. Using a six-row planter, 30 acres of tomatoes can be planted in a day.[22]

The treatment of seeds with fungicides and insecticides before planting to protect against soil disease and pests is a standard practice. Research is being done to treat seeds with a mixture of pesticides, including insecticides that will be taken up into the plant to protect it as it gets older. Also, fertilizers (especially trace elements), growth regulators, and other chemicals may be added. Experiments with seeds of alfalfa, corn, and soybeans treated with herbicides show promise in controlling weeds that compete with young seedlings, with less poison being needed per acre.[23]

Mulches. An old idea from the 1930's has been resurrected and modernized to reduce soil erosion and increase soil organic matter in vegetable fields. It is interplanting, or a "living mulch." Tests by Dr. Robert Sweet of Cornell University found that planting alfalfa, or certain varieties of clover or grasses, about five weeks after the vegetables were planted gave yields comparable with conventional methods and at the same time reduced wind and water erosion, and improved soil structure and organic matter. To keep the living mulch from crowding or competing with the vegetable crop, it has to be mowed or sprayed with small amounts of herbicide. Some additional nitrogen is made available when a legume mulch is used, but the fibrous roots of grasses add more organic matter. Said Dr. Sweet, "Soil erosion and leaching are problems that haven't been addressed by the vegetable industry."[24]

The use of black plastic sheet mulch in vegetable growing has been standard practice for some years. The black plastic absorbs the sun's heat and gets plants off to a quick start, retains soil moisture, and controls weeds. Some plastic mulches are white, which allows plants to capture the sunlight reflected by the mulch and increases photosynthesis,

[21]*Ibid.*, December 1981, p. 9–10, 12; December 1982, p. 26, 28.
[22]*National Geographic*, 1970, vol. 137, no. 2, p. 153.
[23]*Agriculture 2000, A Look at the Future*, 1983, p. 100–101.
[24]*American Vegetable Grower*, September 1981, p. 38, 40, 42, 44, 46, 61.

with a nearly 20% increase in fresh weight of tomatoes in one test. The Japanese have experimented with a clear plastic sheet coated with one or two herbicides to kill emerging weeds for up to three weeks. The use of clear plastic sheeting to *cover* seedlings to form a miniature greenhouse (tunnel covers) is also becoming a common practice.[25]

Bag culture. Another innovation in vegetable growing is called bag culture. Plastic or woven bags about 42 inches long and 15 inches wide are filled with sterilized artificial soil mix (peat-vermiculite) and laid out flat in rows on a greenhouse floor. Holes are cut or burned in the tops (actually the broad sides) of the bags and seeds or seedlings are planted. Smaller upright open-topped bags are sometimes used. The plants are watered by an automatic drip irrigation system, occasionally are given doses of liquid fertilizer, and are treated with whatever pesticides are needed to prevent problems. Dark colored bags absorb heat in cool weather. Yields are excellent, and the bags and pre-sterilized "soil" reduce weeds and soil-borne diseases. All sorts of vegetables and flowers have been grown successfully with bag culture. Bags can be re-used three or four times before the "soil" must be resterilized, saving money.[26]

Irrigation. Agriculture uses over 80% of the water consumed in the U. S., nearly all for crop irrigation. Parts of this country and the rest of the world are in a water crisis—or soon will be. The most critical areas in the U. S. are Arizona, California, Texas, and New Jersey. The availability of clean, fresh (non-salty) water is essential for growing crops. In dry climates, rainfall does not supply enough, streams may be inadequate or already being nearly pumped dry, and underground reservoirs of water are fast diminishing. Much water is too saline (salty) to grow many crops. New or more efficient methods of obtaining water and irrigating are being developed, along with drought and salt-tolerant crops, mulching, better weed control (weeds take water that crops could use), and chemical leaf sprays that reduce the plants' loss of water through their leaves (anti-transpirant sprays).

The traditional methods of irrigation are very wasteful of water. Canal or ditch irrigation has large losses from evaporation, as does sprinkler irrigation. Methods to improve the efficiency of these systems include getting fields perfectly level with the aid of a laser beam in the middle of the field that is detected by a receiver on a leveling machine. A hydraulic control system automatically adjusts the height and angle of the blade. The water is then distributed evenly to all plants. Release of water into a ditch system in intermittent surges by opening and closing

[25] *Ibid.*, October 1981, p. 14, 20–21; June 1982, p. 40.

[26] *Ibid.*, November 1981, p. 15, 34, 36; March 1981, p. 36–37; November 1982, p. 42, 44.

Modern spray irrigation rig and equipment.

valves (called surge-flow irrigation) allows water to reach the far ends of the field more efficiently, since previously wet soil tends to shed water onto dry areas. As much as one-half the water is needed.[27]

The use of lower water pressure on sprinkler systems saves water and still provides adequate irrigation. A new type of sprinkler, the linear-move system, automatically connects and disconnects from supply pipes about every 50 feet. Still another new system is called cablegation. It uses a large plastic pipe with holes drilled at intervals along its sides (called a gated pipe). As water is released into the pipe, a plug attached to a cable slowly travels ahead of the water. As the plug passes one of the holes, a maximum flow of water is released, with less water and then none being released as the plug advances.[28]

New methods to determine the best time to irrigate can reduce water waste. A hand-held infrared thermometer is pointed at plants; it reads their temperature. If their temperature is lower than the air temperature, they have enough water, since their leaves are cooled by evaporation of water through their leaf pores. If their temperature is above air temperature, irrigation should be done. Measurement of soil moisture by an instrument called a neutron-probe is another way to judge the need for irrigation. A space-age method is the use of remote-sensing from an airplane or satellite. The entire system can be computerized to

[27] *Ibid.*, February 1982, p. 6, 44–45; *The Grower*, August 1982, p. 8–10.
[28] *The Grower*, February 1982, p. 29.

control pumps and valves, based on a pre-programmed schedule or weather and soil data.[29]

For fruit and vegetable crops more so than grains, trickle or drip irrigation systems can save large amounts of water. Instead of drenching all the soil, precise amounts of water are delivered only to the root zone, allowing half as much water to be used. A system of plastic tubes or pipes delivers water to small open tubes or nozzels called emitters. The system can be buried below ground or be suspended, as in greenhouses. Usually a plastic sheet mulch is also used to reduce evaporation and weeds. More plants per acre can be grown than with conventional methods. Liquid fertilizers and pesticides can also be distributed through the system, and the operation is usually automated or computerized, greatly reducing labor costs. Some problems are that plastic tubing deteriorates after a couple years and must be replaced, and buried emitters tend to get clogged by soil particles, algae, or mineral deposits.[30]

No-soil agriculture. Growing plants without ordinary soil has been around for a long time. Hydroponics was first developed in the 1930's and was used during World War II to grow fresh vegetables. Some amazing methods are now being developed for the growth of vegetables, greenhouse plants, and tree seedlings without soil. Expenses are high, but for high value crops, such methods are feasible. Yields up to 100 times greater than field production have been obtained for lettuce.

Hydroponics. Growing plants in a water solution of nutrients is called hydroponics. Sometimes plants are grown in a supporting bed of sand, gravel, peat, or vermiculite. Some early methods used large pools or troughs of nutrient solution, but modern methods, called nutrient film technology, save much water by using narrow plastic troughs through which a thin film of nutrient solution is continuously pumped (and recycled). Plastic pipes and small tubes distribute the solution, which can be continuously sterilized to control root disease pathogens.[31] The latest development spaces troughs close together for young plants and farther apart as plants mature, thus saving valuable greenhouse space. Still another futuristic method uses flexible coiled plastic strips ("tape") that are slowly unwound and pulled down troughs of nutrient solution. Seedlings that were germinated in small plastic cups ("cells"), with slits in the bottom for roots, are "planted" in the moving tape at one end. They grow as they slowly move along and when they reach the other end in 20 to 40

[29]*Ibid.*, March 1982, p. 12–13; August 1982, p. 8–10.

[30]*Farm Journal*, December 1981, p. 36–37; *American Vegetable Grower*, November 1981, p. 97.

[31]H. Kahn, *et al.*, *The Next 200 Years*, 1976, p. 127–131.

days, they are ready to harvest. Vegetables can be grown at two or three levels or tiers in a greenhouse, making more efficient use of space. At current prices, lettuce can be grown for two cents per head, compared to seven to ten cents for field-grown lettuce.[32]

Two or three different crops can be grown at different heights in a greenhouse. In an exhibit at the EPCOT Center in Florida, leafy vegetables are grown through holes in styrofoam boards floating in pools of water (with fish also being raised in the water), while climbing crops such as cucumbers or melons are grown on vertical frames.[33]

Aeroponics. Still more futuristic methods are being developed, with potential for growing crops in space. Instead of growing plants in water solution, they are grown in humid chambers on styrofoam boards with their roots being sprayed with a nutrient mist on a computer-controlled schedule. Yields up to 200% higher than soil growth have been obtained, and plants are said to be healthier because they do not have to fight soil-borne diseases. Much less water, fertilizers, and energy are used than with soil farming or even hydroponics. Greenhouse space can be used efficiently by growing plants on several tiers or on slanting "A-frame" styrofoam boards. Calculations indicate that 1/4 acre of aeroponic tomatoes can produce over 100 tons of fruit, compared to 20 to 30 tons for ordinary greenhouse culture.[34]

In the EPCOT Center, vegetables are grown on the walls of an 8-foot revolving drum, which could be developed for use in a space colony. Fluorescent lights in the center of the drum keep plants growing toward the center.[35] A similar system developed by Batelle Pacific Northwest Laboratories uses a revolving cylinder holding the plants and nutrient solution in the center of the growth chamber, with fluorescent lights around the outside, in an opposite arrangement from EPCOT's.[36]

Pest and weed control. Monsanto Chemical Company has developed varieties of alfalfa that are resistant to the commonly-used herbicide "Roundup." Laboratory cultures of alfalfa cells were exposed to lethal doses of "Roundup." Cells that survived were regenerated into whole plants. This development will allow more effective weed control in alfalfa fields.[37]

Rather than indiscriminantly spraying the countryside with extremely toxic chemical pesticides, scientists are researching better meth-

[32]*American Vegetable Grower*, December 1982, p. 30, 33.
[33]*Farm Journal*, February 1982, p. B-1–B-2.
[34]*American Vegetable Grower*, November 1981, p. 102.
[35]*The Grower*, November 1982, p. 24, 28.
[36]*Agriculture 2000, A Look at the Future*, 1983, p. 111–113.
[37]J. W. Rosenblum, *ibid.*, p. 179.

ods of weed and pest control. To reduce the amount of insecticide sprayed on vegetable cropland, a photoelectrically operated sprayer has been developed. An infrared light beam is shone which triggers the sprayer only when it passes over the rows of plants. That way, no insecticide is sprayed on the ground; less than half as much insecticide was required for cabbages, with insect control just as good as with conventional spraying.[38] New low-volume pesticide sprayers (pulse-jet foggers and aerosol generators) reduce the amount of pesticide sprayed but give good pest control in greenhouses if the proper pesticides are used.[39]

Another approach widely used in recent years to reduce pesticide spraying is to only spray when it is needed, rather than "just in case" (as a preventative). Frequent and careful field scouting of crops is necessary, but many dollars of chemicals can be saved, plus the cost of spraying and potential ecological damage. Recently an "electronic field scouter" was developed to monitor apple orchards for apple scab disease. It is a small box (one foot high) containing a computer microprocessor. A keyboard allows the operator to punch in data on the past history of apple scab and the weather. Current weather conditions are monitored from sensors in the orchard. The instrument then considers the relevant factors and tells the operator whether he needs to spray fungicide or whether he can skip a spraying (and save money).[40] The same sort of electronic blight forecaster has already been used by potato growers. Similar systems are being developed for other diseases as well as insect pests (powdery mildew, the codling moth, nematodes, weeds).[41] A hand-held infrared "gun" thermometer allows detection of brown stem rot in soybeans, because stressed plants have a slightly higher temperature than healthy ones.[42] A still newer method, not yet ready for field use, uses the faint light emitted from chlorophyll that has been exposed to light ("delayed light emission") to detect stress in leaves before other stress symptoms become visible.[43]

Another method of reducing pesticide use is integrated pest management (IPM). In this common-sense approach, biological control methods are used as a "first line of defense." These would include resistant varieties of crop plants, predaceous insects such as ladybugs and green lacewings, weed-eating insects, parasites of pests such as tiny parasitic wasps, diseases of insects and weeds, the release of sterilized insect

[38]*American Vegetable Grower*, August 1982, p. 25.
[39]*Ibid.*, November 1982, p. 28, 30.
[40]*The Grower*, February 1982, p. 50–51.
[41]*Ibid.*, January 1982, p. 54–55.
[42]*Science*, 1979, vol. 205, p. 585–587; *Successful Farming*, December 1981, p. 26–P.
[43]*Science*, 1982, vol. 215, p. 1104–1106.

pests in order to reduce their reproductive rate, use of natural insect repellents as well as attractants (to poisoned baits or traps), and behavior-modifying chemicals to interfere with normal feeding or reproduction. Only if weeds or pests threaten to devastate a crop are toxic chemicals used, and then only as needed. Again, careful field scouting is required to monitor crops.

Recent research at the University of Illinois has discovered a way of causing weeds to "commit suicide." All cells contain very small amounts of a certain amino acid, abbreviated ALA (delta-aminolevulinic acid). When larger amounts of ALA extracted from a certain species of weed are sprayed on plants in the afternoon, the next morning's sunlight will stimulate the weed's tissues to disintegrate in a few hours. This happens because the ALA is converted into another substance (tetrapyrroles) which absorbs light and releases high amounts of activated oxygen, which kills cells. The ALA and tetrapyrroles decompose within five hours, and they may be able to be manufactured by genetic engineering techniques.[44]

Remote sensing. To provide for better crop monitoring as well as other surveying uses, the space-age technology of remote sensing is gaining prominence. Remote sensing involves recording and analyzing various natural radiation from the earth's surface, such as visible light, infrared, and microwave. Pictures taken from aircraft or satellites can be computer-analyzed and reveal an amazing number of things about the earth, such as weather patterns, ocean currents and icebergs, geological features and potential mineral resources, data for mapping and land-use planning, and numerous agricultural applications. These include mapping cropland use and soil types, inventorying crops and livestock over large areas, crop yield estimates, and monitoring crops for pests and diseases. By using infrared pictures, sick plants can be told from healthy ones. Large acreage can be monitored quickly and the need for spraying pesticides determined. Amazing detail can be seen using extremely fine-grained films.[45]

Animal agriculture. Just as in crop technology, animal agriculture is undergoing a revolution. For a number of years there has been an increasing trend toward larger, centralized animal-raising units, with more automation. These trends are expected to continue. Aquaculture, the growing of aquatic plants, fish, shrimp, lobsters, oysters, clams, etc., is expected to provide a needed source of protein for a future hungry world. Some of the recent developments in typical land livestock and

[44] *Science News*, September29, 1984, p. 197; *Acres U. S. A.*, January 1985, p. 4.
[45] *Science*, 1980, vol. 208, p. 670–679; 1981, vol. 214, p. 139–45.

poultry raising include genetic improvement and increased reproduction, better nutrition and growth, and new confinement and management methods.

Genetic and reproductive research. Animal breeders have long tried to breed improved strains of domestic animals, whether dogs, rabbits, pigeons, or livestock. Cross breeding, computerized record keeping, and artificial insemination have already speeded up the development of improved breeds. More recent research includes use of synthetic hormones to synchronize the estrus cycle of cattle, sheep, and swine in order to achieve high rates of conception; causing superior animals to release large numbers of eggs and transferring embryos to other females for pregnancy; freezing embryos for future growth or for transport to faraway places; cloning embryos from superior animals; controlling the sex of offspring; and genetic engineering to produce greatly improved animals (producing more milk, meat, eggs, or wool; better quality, such as more meat and less fat for a carcass; giant or dwarf varieties; and disease resistance or immunity) and to produce vaccines and drugs for animal diseases.

In swine, the performance of different boars can be evaluated by heterospermic comparison, mixing semen from two boars and inseminating one sow. When the litter is born, it can be found which boar sired more and better offspring.

Nutrition and growth. Researchers are always looking for ways to raise animals more efficiently and cheaply. Disease and poor health currently account for an 11% loss of livestock per year. Faster growth on less food may mean the difference between profit and bankruptcy. Some livestock, such as sheep and beef cattle, can make use of low-quality rangeland vegetation, on land not suitable for crops. For confined animals, diets and feed rations are highly refined to insure that the animals receive adequate amounts of protein, vitamins, and minerals. Administration of growth hormones or stimulating the body to produce more growth hormones improves feed efficiency and increases rate of gain and amount of meat. Milk production could jump 15 to 40%.

Cheaper sources of feed, plus recycling of wastes can greatly reduce the cost of production. Some examples include feeding poultry vegetable instead of animal protein; use of by-products such as distiller's grains, corn gluten meal, potato wastes, supermarket vegetable clippings and unsalable produce, wood wastes, and elevator grain dust; use of recycled animal wastes, such as poultry litter and various animal manures, which contain a large percentage of undigested food; and plastic pellets for ruminant animals, which provide needed roughage and scratchiness (these can be recovered, cleaned, and recycled). Meat animals can be fed these cheap foods and then be finished in a feedlot on hay, silage, or grain.

Management technology. Most livestock can be raised most efficiently in a confinement system. Cattle feedlots and dairies, swine farrowing houses, and broiler and "egg factories" are familiar today. More mechanization and automation will occur. Controlled environment systems will regulate temperature, humidity, ventilation, feeding, and manure-removal. Filters will remove disease pathogens from the air.

Automatic metering devices will dispense feed and drinking water, along with drugs, tranquilizers, or growth hormones. For poultry, slow-moving conveyors will move caged hens along rows as they mature; at the end of the line, the conveyor loads them onto trucks for transport to a processing plant. For larger animals, a small transponder can be put in a collar or leg band, be lodged in the rumen, or be implanted under the skin. When the transponder receives a radio signal, it identifies that animal. The system can be used to dispense the proper amount of feed for that individual, as well as to keep a computerized record of each animal's milk production or state of health or estrus (by reading implanted sensors). Automatic manure flushing systems will remove manure to methane production units or water-removal processors before being re-fed to the animals. Recovering ammonia from manure and converting it into fertilizer is another possibility.

Information gathering/processing. This is the computer age. The things these man-made marvels can do are truly amazing. From room-filling monsters only a few decades ago to today's micro-computers you can hold in your hand, progress in computer technology has been incredible. Just from 1970 to 1980, the speed of computers increased 100 times, the number of electronic components per square inch increased 10,000 times, and the cost of use decreased 90%. Future developments may be as amazing. The use of fiber optics instead of metal wires may allow computations at the speed of light, and memories using special types of molecules may allow super-memories or micro-miniature computers. Computers will be commanded by voice as well as buttons, and they will answer by voice.

The computer is just beginning to be used in agriculture. Presently, few farms have computers and relatively few agriculture-adapted programs have been written (because of the small demand). Growth along these lines will accelerate in the future.

Computers will become easier to use (more "user-friendly"). Programs (software) will be written for every aspect of farming: planning which crops to plant; how much and which fertilizers to buy; projected yields, depending on weather conditions; record keeping of crop histories, animal production, fertilizer and fuel costs, etc.; accounting, tax computation, and market analysis.

Central data banks and computer networks will allow an isolated farmer to receive information on a multitude of subjects, such as crop fertilizer requirements, climate data, or the current status of the wheat crop in Argentina (via satellite remote sensing). The farmer can instantly communicate with other computers in the network, buy supplies, sell his commodities, do his banking, get the latest weather forecast, or ask the extension agent for advice.

As mentioned earlier in this chapter, computers, microprocessors, sensors, and other electronic marvels will do such things as steer tractors, adjust combine cutting, monitor soil and crop conditions, keep records of individual animals, run irrigation systems and hydroponic greenhouses, selectively spray weeds and not crops, detect subclinical mastitis in cows, detect diseased crops, and so on.

Will it happen?? We truly live in an age of man-made miracles. Modern science and technology can do almost anything they want. They have opened a bottle with a genie inside.

Will this vision of the future come to pass? Is this what we want agriculture to become? Are there any other ideas on what the future will bring?

The rosy, whiz-bang, fantasy land painted by the futurists is only one possibility. Now let's look at the opposite extreme.

The limits to growth. Trends that became apparent a few decades ago troubled certain economists, sociologists, and biologists. In 1972, a book called *The Limits to Growth* was published for the Club of Rome, a group of 30 (at that time) scientists, educators, humanists, industrialists, and civil servants from ten countries. They met in 1968 to study the "predicament of mankind," the complex problems facing all mankind: poverty in the midst of plenty, degradation of the environment, the population explosion, uncontrolled urban sprawl, rejection of traditional values, and economic troubles. Their book, *The Limits to Growth*, became an international best seller. It envisioned an entirely different future from what we have just seen.

The Club of Rome and a team of researchers at the Massachusetts Institute of Technology used *models* (mathematical descriptions of how various processes change in time) of five main factors that control growth of human activities: population increase, agricultural production, nonrenewable resource depletion, industrial output, and pollution generation. Each of these major factors was subdivided into sub-factors that control it, with about 120 factors included in their "world model."

For example, in the area of agriculture, food production is increased by increasing the amount of cultivated land and capital investment (machinery, fertilizers, irrigation systems, etc.), but is decreased by soil losses and pollution. Food production helps determine the population size, along with birth and death rates. When we examine how various factors have changed in the past, we find that most factors increase (or decrease) *exponentially*. That is, instead of increasing at a constant rate, such as a charitable fund receiving $100 every month, things increase at an ever increasing rate. This exponential growth occurs in a bank savings account earning interest, say at a rate of 8% per year. A beginning amount of $100 would become $108 the first year, $116.64 the second year, $125.97 the third year, and so on. At first the growth is slow, but in

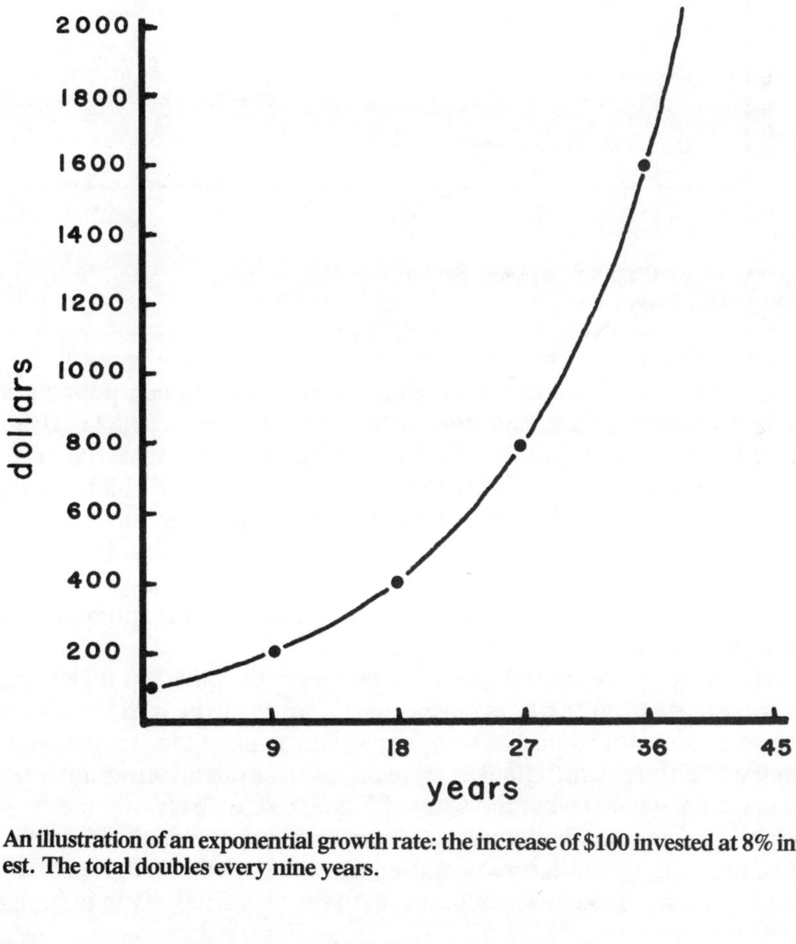

An illustration of an exponential growth rate: the increase of $100 invested at 8% interest. The total doubles every nine years.

9 years, the $100 would nearly double, and would double again in the next 9 years. It only takes 14 doublings to make you a millionaire!

Simple organisms such as yeast and bacteria divide into two new cells every 20 minutes, making huge populations possible in a short time. Most organisms reproduce at a lower rate, but unless checked by factors of their death rate, they would soon overpopulate their environments. Exponential growth rates cannot continue indefinitely unchecked; "something has to give."

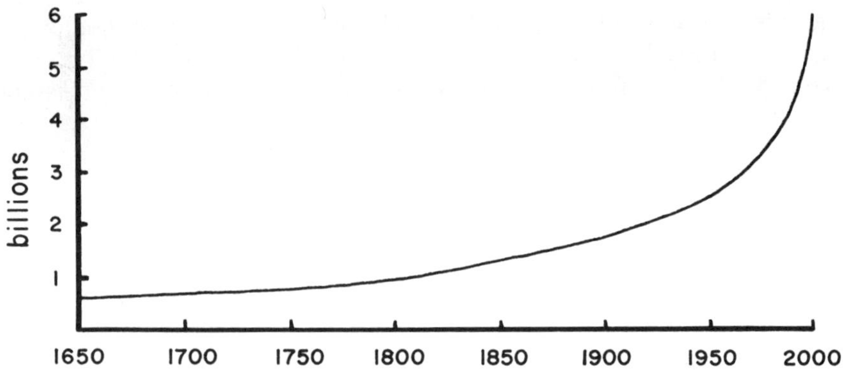

Exponential growth of world population since 1650 (Source, D. J. Bogue, *Principles of Demography*, 1969).

Growth rates of natural and human phenomena are controlled by positive and negative influences. For example, the human population level is increased by a higher birth rate and decreased by a higher death rate. In turn, the birth rate is influenced by the population size (the more people, the more babies born), percent of population of child-bearing age, food supply and affluence of the society, the lag time between birth and maturity, and the culture's desire to have large families. Death rate also has several influences. So we see that all of the many parts of nature and society are *interconnnected*; one affects another, and often in complicated ways, since several factors are involved.

By studying the rate of growth of factors in the past, it is then possible to project what might occur in the future if the growth rate is unchanged, and with the aid of computers, it is possible to see what might happen at different and changing growth rates, as other factors influence the first one. That is what the Club of Rome did, with a world model of about 120 factors. Their research team fed various numbers, growth rates, and theoretical future conditions into the computer and got various projections of what might occur up to the year 2100. Even though it contains about 120 factors, the Club of Rome researchers emphasize that the model is still oversimplified and does not describe the real world with great accuracy. Thus, the projections made using it are not intended to be accurate as to the exact times and quantities, but *they are accurate enough to show general trends*. They are not really predictions, but possibilities.

When the researchers ran the data through their computer with the assumption that future conditions, rates of growth, and policies will be the same as they are today, the results were spine-chilling! The computer indicated that because of soaring world population, increasing pollution, and natural resource depletion (minerals, oil, fertilizers, etc.), industrial production, food production, and then world population will come crashing down to disastrously low levels (to about the levels of the year 1900) *within the next 100 years*. Even if the assumption is made that twice as many natural resources would be available as we now know exist (from unexpected new finds of minerals), the crash would still happen, only it would be delayed a few decades. The same thing holds true even if we assume (it is easy to make these assumptions, change the numbers, and have the computer re-run the data) that natural resources are unlimited (pollution would be the cause of the crash), or that pollution is also controlled (then population and industry expand and use so much farmland that food production is hurt), or that agricultural productivity is also doubled (then increased population and industry overload the world with pollution), or even that "perfect" birth control is also attained (then industrial growth slows and resources are depleted as pollution and food shortages choke off population growth). The more changes we make to improve things (pollution control, birth control, more food, etc.), the greater the extremes of growth and resulting crash. This is because at the upper levels of exponential growth, tremendous increases take place within only a few years. Also, there are unavoidable time lags between many causes and their ultimate effects, such as the 15 or so years between birth and reproductive maturity in humans, and the approximately 20 years before many environmental pollutants cause cancer. It takes time for changes in government policies to be made and to take effect. When growth rates are slow these lags are not so important, but when exponential growth is doubling perhaps every seven years, there is not enough time to effectively deal with the problems. The Club of Rome's world model also does not take into account such possibilities as political upheavals or wars which could disrupt industrial and agricultural production.

The earth in its present state is finite, a closed system. It cannot supply enough resources and land, produce enough food, and absorb enough pollution for the world's human population to grow indefinitely. Already we are seeing awesome problems with no apparent solutions. In third world countries, burgeoning populations with little income flock to already overcrowded, polluted cities. The "quality of life" in most affluent countries' cities isn't so great either. Crime, corruption, breakdown of the family, and immorality are increasing. The known world reserves of aluminum will last only 31 years at exponential use rates, copper supplies will only last 21 years, mercury and silver 13 years, and gold only 9 years.[46] Many long-lasting pollutants (DDT, lead, mercury) are found world-wide in the environment and food chain. Even the oceans and groundwater supply are fouled. The forests and jungles of the tropics are being cut for lumber, firewood, and land clearing at such a great rate (27 million acres a year, or 50 acres a minute), that biologists are fearful that many potentially useful species of plants and animals will soon become extinct.[47] Deserts are expanding at the rate of 23,000 square miles a year.[48]

The rate of increase of carbon dioxide in the atmosphere from industry, automobiles, volcanoes, and burning of wood by poor cultures has reached the point that some scientists believe irrevocable changes in world climate will soon occur. Some think the high carbon dioxide will increase the atmosphere's "greenhouse effect," with resulting melting of the icecaps and flooding of coastal cities.[49] Others believe that increasing carbon dioxide coming from volcanoes will throw the earth's climate into an ice age cycle of glaciation. Volcanic dust in the atmosphere reflects part of the sun's light, causing climate to cool. Increased carbon dioxide does trap more of the sun's heat by the greenhouse effect, but then more heat is radiated back into space at night and especially at the poles in the winter. Hot air in the tropics rises and flows toward the poles, allowing colder northern air to flow toward the equator, cooling the climate in the temperate zones. More moisture is carried toward the poles and dumped as snow, beginning glacial build-up.[50] Climate expert Reid Bryson believes the cooling of the earth's climate that has occurred since 1940 is due to increased dust and other particles in the atmosphere as a result of increased industrialization, urbanization (automobiles, smog), and volcanic activity. Sunlight is blocked, overpowering the warming caused by the carbon dioxide greenhouse effect.[51] World

[46]*The Limits to Growth*, 1972, p. 64–66.
[47]*Science*, 1983, vol. 221, p. 1168–1169.
[48]R. Allen, *How to Save the World*, 1980, p. 12–13.
[49]*Science*, 1983, vol. 222, p. 491, 1081–1086, 1107–1108.
[50]J. D. Hamaker, *The Survival of Civilization*, 1982, p. 69–72, 148–158.
[51]*Weatherwise*, 1968, vol. 21, no. 2, p. 56–62.

weather has been increasingly upset, especially since 1972, with millions of dollars of damage caused by each blizzard, flood, or hurricane.

Still another problem is the rapid rise of methane gas in the atmosphere, about 1–2% per year. Like carbon dioxide, it can disturb the earth's climate by the greenhouse effect and influence the production of smog.[52]

In *The Limits to Growth*, the Club of Rome researchers commented on the optimistic hope of many people that new technological developments will somehow allow us to avoid the projected crash. They say that (1) technological change requires a lag time to become accepted by people and implemented (such as the "Green Revolution" in third world countries, where there may be cultural resistance to change and lack of money to implement), (2) technology may only temporarily relieve the symptoms of a problem without attacking the real causes, and (3) many problems have no technological solution, such as racial tensions, unemployment, and the nuclear arms race, but require a change in human values or morality (p. 152–160).

Reactions to the Club of Rome's book, *The Limits to Growth*, were widely divergent. Some hailed it as the handwriting on the wall. It was published in the midst of a period of ecological awareness and activism and found a ready audience. But others were very critical of it. In 1973, a book called *Models of Doom: A Critique of The Limits to Growth* was published, followed by *On Growth* in 1974 and *On Growth II* in 1975. These books consist of essays and interviews with scientists, historians, philosophers, statesmen, and economists. Some of them agreed with the Club of Rome to some extent, but others did not. The Club of Rome was accused of being prophets of doom. Their study was attacked for not believing that technological advances could overcome pollution and other problems. Their computer world model was criticized. It did not consider economic and social factors nor technological advances. Their data were said to be incomplete or unrealistic.

Was the bearer of bad tidings being attacked because we didn't like his message? Are we prone to ignore unpleasant news and stick our heads in the sand? Was the Club of Rome's study a complete surprise?

Other warnings. The Club of Rome's study was not the first warning that mankind may be heading toward ecological disaster. Way back in 1798, in his *Essay on the Principles of Population*, Thomas R. Malthus stated that population increases geometrically (exponentially), but the food supply only increases arithmetically (at a constant rate). Thus, human population will increase until checked by famine, disease, or war.

[52]*Science*, 1984, vol. 226, p. 33, 954–955.

Malthus' numerical assumptions are no longer believed accurate, but his warning was far ahead of its time.

Charles Darwin made use of Malthus' ideas in developing his theory of evolution, but few people became concerned about ecological problems until after World War II, when it began to be obvious that man's numbers and activities are indeed causing serious problems. Fairfield Osborn, in his books, *Our Plundered Planet* (1948) and *The Limits of the Earth* (1953), emphasized that we are all in this together, all nations and peoples share a common earth, and its resources are fast disappearing because of man's misuse of them.

In 1962, Rachel Carson's blockbuster, *Silent Spring*, told us about the ecological danger of accumulation of persistent pesticides (DDT) in the food chain. *Famine 1975!* by William and Paul Paddock, published in 1967, forcefully pointed out that a population-food collision is inevitable and predicted widespread famine by 1975. The authors showed how such technological developments as increased fertilizer use and irrigation of new lands, new hybrids, harvesting the oceans, and growing algae simply cannot avert the coming disaster. The U. S. will soon run out of food surpluses. They believe the U. S. will have to make the cold blooded decision to allow the worst-off nations to starve and help those who may still make it.

The Population Bomb, by Paul Ehrlich (1968) sounded the alarm about the impending population explosion and famines. The rate of increase in world population is rapidly increasing, with the time required for the population to double decreasing from about 1000 years in the Middle Ages to 200 years in the 18th and 19th centuries to 80 years from about 1850 to 1930, to only about 35 years today. At least half of the world is already undernourished or malnourished. One out of three people may starve before long. Because of pollution and environmental degradation, the earth is a dying planet.

Many other books, studies, and reports appeared during the late 1960's and 1970's. Books include *Starvation or Plenty?* (1970), *The Passing of the Modern Age* (1970), *Only One Earth* (1972), *The End of the American Future* (1973), *The Coming Dark Age* (1973), *The End of Affluence* (1974), *The World is Coming to an End!* (1975), and *The Next 200 Years* (1976). Studies include *Growth of Population* (Proceedings of the First Conference on Population, Princeton, New Jersey, 1969), *Man's Impact on the Global Environment* (Report of the Study of Critical Environmental Problems, MIT, 1970), *Environmental Quality* (Second Annual Report of the Council on Environmental Quality, Washington, D. C., 1971), and *Problems of the Human Environment* (Report of the Secretary-General, United Nations, 1971).

In 1977, President Carter got into the act and directed the Council on Environmental Quality and the Department of State to study the "probable changes in the world's population, natural resources, and environment through the end of the century." The result, called *The Global 2000 Report to the President* (1980), contained "disturbing" conclusions, showing the potential for "alarming" global problems of the environment, resources, and population by the year 2000. In a summary of major findings and conclusions, the report's authors state, "If present trends continue, the world in 2000 will be more crowded, more polluted, less stable ecologically, and more vulnerable to disruption than the world we live in now. . . Despite greater material output, the world's people will be poorer in many ways than they are today. . . Indeed, the problems of preserving the carrying capacity of the earth and sustaining the possibility of a decent life for the human beings that inhabit it are enormous and close upon us."

The report singled out the serious problems of world food production, energy shortage, water shortage, loss of forests, deterioration of agricultural soils (due to erosion, compaction, loss of organic matter, spread of deserts, increased salinity and alkalinity, loss to urbanization, and waterlogging), increasing atmospheric carbon dioxide, and extinction of plant and animal species. Deteriorating soil and agricultural resources were called "perhaps the most serious environmental development," especially in lesser developed countries. The report's study team assumed that present public policies would in general be continued, that rapid technological development would continue, and that there will be no major disruptions of international trade from wars or political and economic difficulties.

In a recent news release (September 20, 1984), nineteen Nobel Prize scientists and 178 leaders of environmental and arms reduction groups said, after a conference on "The Fate of the Earth," that "unless humanity changes its ways, mankind faces extinction either through a nuclear war or an environmental catastrophe." As others have done recently, the group warns that even a limited nuclear war could produce enough smoke from burning cities to block out nearly all of the sunlight in the northern hemisphere, plunging the earth into a dark, cold "nuclear winter."[53] Ordinary agriculture would be impossible. More recently, these dire warnings were toned down.[54]

A recent year-long study by the World Watch Institute concluded that nearly half the world's cropland is being eroded at such a rate that

[53]*Ibid.*, 1983, vol. 222, p. 822–823.
[54]AP News Release, January 10, 1985.

our long-term ability to grow food is being damaged. Already erosion by wind and water is contributing to famine in Africa. Soil loss in the U. S. is comparable to the 1930's Dust Bowl, with 44% of the land eroding faster than topsoil can be replaced.[55]

After their original book, *The Limits to Growth*, the Club of Rome researchers, as well as other study groups, used other more refined computerized "world models" to try to get a more detailed look at the future. A later model of the Club of Rome divided the world into ten regions and analyzed their growth and the problems of developed and undeveloped economies separately.[56] The conclusions still were that undirected and uncontrolled growth will cause severe problems in some regions, which would be felt by the whole world. Anything less than full international cooperation would be disastrous. Projections include a long-term trend of steeply rising food prices and massive famine in Asia and Africa before the turn of the century.

The system at present. How well is "the system" working today? How effective are world governments at dealing with and preventing problems? Are national economies and industries providing an adequate standard of living for their people? Is agriculture feeding everyone with enough healthful food? Are modern science and technology creating the good life?

We have described some of today's problems earlier in this chapter and in Chapter 1. A few more examples.

Green Revolution. When it first appeared in the mid-1960's, the "Green Revolution" was hailed as the answer to world hunger. Its architect, Norman Borlaug, at first said that just by using and improving on conventional food producing methods, plus developing newer nonconventional methods, "man can feed the world's mushrooming human population for the next 100–200 years."[57] But the revolution is turning from green to brown. The "miracle" varieties of wheat and rice that Borlaug developed require expensive inputs of fertilizer, pesticides, irrigation, and mechanization. Rich third-world landowners adopted them, but the majority of poverty-stricken farmers were left out. The new varieties of grain, developed in temperate climate, and the technology were not well adapted to the hot tropics. The local people did not like the grains' taste. Planting many acres to a single variety (monocropping) increased the danger of wildfire spread of diseases and pests. Yields were indeed greatly increased (2 or 3 times the local varieties), which caused a

[55] *Wisconsin State Journal*, September 30, 1984.
[56] *Mankind at the Turning Point*, 1974.
[57] *Phytopathology*, 1965, vol. 55, p. 1088–1097.

glut on the local markets and low market prices. Many farmers who tried the "miracle" grains stopped growing them because of low prices, low quality of the grain, and added expenses. On non-irrigated land, yields were no better than local varieties. Very little land in the world is irrigated or can be (at great expense). The increases in world food supply that occurred between 1967 to 1969, can be traced to the favorable weather during those years, not to "miracle" grains, since the yields of other traditional crops increased as well as the new varieties. William C. Paddock calls the Green Revolution a disaster if there is no population control to go along with it, for more food means that more people will live, accelerating the population explosion.[58] Even Norman Borlaug later shortened the time that the Green Revolution can feed the world from 100–200 years to "two or three decades."[59] For a hard-hitting indictment of government and land-grant college research orientation, read *Hard Tomatoes, Hard Times*, by Jim Hightower (1972).

Farm policies. How about the government's efforts to make things better? How are they working? Look at some recent newspaper and magazine articles. "Farm Income Level Lowest in 12 Years," reads an AP news release of July 5, 1984. "Government 'Help' " has Hurt Farming," reads another (May 2, 1984). It states simply that government programs have failed to keep family farms in business, but instead have had the opposite effect: farms are fewer and bigger, and foreign commodity sales are less than they would be without government programs. This is not surprising, since government subsidies encourage farmers to produce more products, resulting in unsold production.

The same things were said in more detail in an article by Purdue University Professor Emeritus Dr. Don Paarlberg.[60] The government commodity programs first begun in 1933 during the Depression failed to save the family farm and improve farm income over the long run (programs were reduced beginning in 1973). In 1930–32, before the programs, farm prices averaged 72% of parity; in 1970–72, they were again 72% of parity. Subsidy programs provided incentives to increase yields per acre; the largest 1% of farmers received 21% of the payments, and used their payments to buy out their smaller neighbors, with the number of farms declining from 6.5 million in 1930 to 2.9 million in 1970. The artificially high market prices led to surpluses and expensive government stockpiles. The programs were costly to consumers and taxpayers; from

[58]W. C. Paddock, How Green is the Green Revolution?, *Bioscience*, 1970, vol. 20, p. 897–902.
[59]*Columbia Journal of World Business*, 1969, vol. 4, p. 1019.
[60]*Wisconsin State Journal*, March 7, 1982, section 5, p. 7.

1964–69, the price of food increased 4% per year over what it otherwise would have been. Government decision-making took away individual freedom of management. Although the programs did help prevent economic disaster in the Depression years, they were continued far too long. "We should read the lesson well, lest we repeat it," said Paarlberg.

In an article titled "Former Ag Secretaries See No Quick Fix for Farm Prices," four former U. S. secretaries of agriculture agreed that farming is in a depression, that there is no instant cure, and that more government programs are the answer.[61]

Pork-barrel projects. The Garrison Diversion Unit is a huge irrigation project in North Dakota. During the dust bowl days of the Depression, the Army Corps of Engineers and the Interior Department constructed a water plan, which later became the Flood Control Act of 1944. The Corps of Engineers built six dams along the Missouri River for flood control. Garrison Dam in central North Dakota was completed in 1953, flooding a half-million acres of land. Some of the water was to be diverted for agricultural use. The original idea was to irrigate a million acres in the dry western part of the state, but soils there were found to be too dense for irrigation.

So a scaled-down plan was formulated to irrigate 250,000 acres in the central and eastern parts of the state. A system of canals and reservoirs is supposed to feed a 25-county area. Strangely, the area is already one of the wettest parts of the state, the "prairie pothole country," with sloughs and marshes and abundant waterfowl. Construction of the first canal, begun in 1970, enraged farmers whose land was torn up. The Canadian government fears the threat of pollution and that undesirable species of fish and fish diseases will make their way into the rivers draining into lakes Winnepeg and Manitoba. The first phase of the project will take 8 to 10 years to complete and will irrigate only 85,000 acres at a cost of over $8000 per acre, thanks to cost over-runs. After that, the cost of irrigation for individual farmers runs about $60,000 for a pump and center-pivot irrigation rig (to cover 160 acres), plus $25 or $30 per acre for operation and maintenance. At current prices, corn is too expensive to raise by irrigation and few farmers would grow alfalfa because of low beef prices, so many feedlot owners are going bankrupt.

Environmentalists and farmers have been trying to kill the project since 1972. Congress has asked the Interior Department to decide its fate. It will take at least two-thirds of the 12-member appointed commission to agree to halt or modify the project. Meanwhile, bulldozers and earthmovers are racing to get as much done as possible before funds

[61]*Farm Journal*, May 1982, p. 24.

might be cut off, so that there will be more likelihood that the project will survive.[62]

Too little, too late? Does it really appear that the political system is seriously trying to solve and prevent the awesome problems facing us? It sounds like "business as usual." True, many of the people who are part of the system honestly believe that they are doing their best to help and that governments and technology will solve man's problems. But are their efforts great enough? Or do the forces causing the problems have so much momentum that no matter what man does, catastrophe is inevitable?

We have seen in this chapter two possible views of the future, two extremes of the experts' opinions. One is unboundedly optimistic; science, technology, and man's good-hearted efforts will save the day and usher in a utopia of prosperity for all. The other extreme is gloom-and-doom, a neo-Malthusian vision of ecological, political, and economic disaster—already begun.

Which will it be??

[62]*Science*, 1984, vol. 225, p. 904–906.

CHAPTER 7

The Future—

The Coming Revolution

"Then I will give you rain in due season, and the land shall yield her increase, and the trees of the field shall yield their fruit. And your threshing shall reach unto the vintage, and the vintage shall reach until the sowing time: and ye shall eat your bread to the full, and dwell in your land safely." Leviticus 26:4–5

Midway between the gloom-and-doom and the super-optimistic views of the future is a middle ground. It is sometimes called controlled growth, sustained growth, zero growth, steady state, or global equilibrium.

Controlled growth. The Club of Rome did not leave us hanging over the fire in their book, *The Limits to Growth*. In that book and in subsequent reports, they gave remedies for the "predicament of mankind." Rather than allowing nature to take its course as a result of uncontrolled growth followed by "overshoot" and collapse, a deliberate, self-imposed limitation to growth can be employed (it won't be easy), resulting in a stable "global equilibrium." The population (controlled by the birth rate) and level of capital investments must be kept constant. This does not mean an unchanging stagnation, but local regions could increase or decrease without changing the overall average. Such activities as education, scientific research, the arts, athletics, and social interactions, which do not use up large amounts of non-renewable resources, could flourish indefinitely. Technological advances in such areas as waste recycling, pollution-free energy, non-toxic pest control, and better product design would be welcome.

Other studies, reports, and books have come up with similar visions of the future, with varying degrees of limited controlled growth. Science and technology would be expected to contribute to ecologically sound expansion and use of resources. The United States, because of its relative wealth and influence, would play a major role in shaping the world's future. It would be assumed that the world's leaders would be altruistic, that people would have high moral strength and be willing to sacrifice for the common good, that there would be no wars or riots, and so on.[1]

Many nations are already embarked on optimistic long-range planning. Some people want to reduce growth to zero and have everyone live a spartan existence, with solar collectors, wood-burning stoves, and a self-sufficiency garden. Others foresee a future of reduced but steady growth and relative affluence for all peoples. The importance of redefining human values and morals is mentioned by some writers. Rather than being individualistic or nationalistic in outlook, people need to see the earth as a whole, a "human community," with concerns about caring for our "spaceship earth" and a sense of responsibility for their activities.[2] We need a realistic long-term goal and the human will to achieve that goal.[3]

But let's get back to agriculture.

Alternatives. Are the ways promoted by the "standard" farm magazines and recommended by the state universities and county extension agents the only way to farm? Are commercial fertilizers, pesticides, and hybrid seed "required" to get decent yields? Are one-fourth of young calves or pigs *expected* to die from disease and birth defects?

As you probably know from reading the earlier parts of this book, not everything about modern agriculture is necessarily good for the soil, nor produces high quality crops or healthy animals. In previous chapters we pointed out some of the problems caused by pesticides and other practices and gave some ways of improving soil and crops; things that were well-known in the past, but have been shoved aside by the glitter of new technology. Even "the establishment" has finally started to recommend some of these things in recent years, such as reducing the amounts of pesticides and nitrogen fertilizer, split applications and in-row application to conserve fertilizer, and recycling organic wastes to build up soil organic matter. "Organic farming practices work," says Robert I. Papendick, USDA soil scientist, "That is no longer a disputed fact among knowledgeable people."[4] Several state universities are teaching courses and doing research on "organic" methods. The big chemical companies are scrambling to do research on soil microorganisms and non-toxic methods of disease and pest control, mainly to develop their own patent-

[1]S. Wortman & R. W. Cumings, Jr., *To Feed this World*, 1978; R. Allen, *How to Save the World*, 1980; E. Goldsmith, *et al.*, *Blueprint for Survival*, 1972; D. Knorr, *Sustainable Food Systems*, 193.

[2]P. Henriot, p. 36–43, in *Making It Happen,* 1982.

[3]*The Limits to Growth*, 1972, p. 188.

[4]*Farm Journal*, March 1982, p. 46.

able strains of microbial inoculants or pest-fighters. They are also buying up smaller companies that already have such products. That should tell you something about the trend of the future.

Yes, there *are* alternatives. In this chapter, let's look especially at some of the lesser known methods and products and at some of the "far-out" ideas being developed, some of which may become the "traditional agriculture" of the future. Get ready to have your mind stretched!

Soil conditioners. Starting with the soil and improving soil structure (tilth), there are a number of materials and products that loosen "dead," compacted soil. They may also help eliminate a hardpan. By loosening soil and allowing air to enter, beneficial soil organisms can flourish and healthier, better quality crops are the result (see Chapter 3). Some of the products do add fertilizer nutrients themselves, but others contain few nutrients and serve only to improve tilth, so are usually called *soil conditioners*.

1. First we have various types of natural mined and ground-up *rock deposits*. They help particles of "tight" soils to cling together to form a desirable "crumb structure" ("good tilth"), plus they may contain significant amounts of secondary and trace elements.

a. *Liming materials*. Ground high-calcium limestone (calcium carbonate) and other high-calcium liming materials (marl, water treatment plant lime, papermill waste, slags) can greatly improve structure in many types of soils, especially eastern acidic, clayey soils. In the western alkaline soils, gypsum (calcium sulfate) may improve soil structure. Certain mined products from the western U. S. that contain large amounts of gypsum shale have been found to improve soil structure and supply trace elements.

Dolomitic lime (calcium magnesium carbonate) is an extremely hard stone and slow to become available to plants. Also, its high magnesium content may provide excessive amounts of this element if it is not needed.

b. *Humates*. Humates are mined near commercial coal deposits in the western U. S. They are called oxidized lignite and are high in salts of humic and fulvic acids. These acids are also found in humus and compost. They have been found to be plant growth stimulants and can chelate trace elements so they are more available to plants. At moderate rates of application (500–750 lbs./acre broadcast), humates can improve soil structure and drainage, rooting, growth, and crop quality, including shelf life of fruits and vegetables.[5] However, too much humic acid will

[5]Tech. Bulletin 1015, South Carolina Agric. Exp. Station, Clemson Univ., 1964; J. C. McAllister, *A Practical Guide to Novel Soil Amendments,* 1983, p. 34, 36; *Acres U. S. A.* June 1984, p. 20–22.

inhibit plant growth, so it is best to keep application rates low (less than 100 lbs./acre broadcast or 15–30 lbs./acre in the row). Humates work best when mixed with fertilizers. They will have little noticeable effect in soil with high organic matter.[6]

c. *Rock phosphates*. There are two common types of rock phosphate products. Ordinary ground rock phosphate, sometimes called hard rock phosphate, is mined mostly for acid treatment to make superphosphate. Some is sold for use by "organic" farmers and gardeners. When the rock phosphate is ground and washed, the washings are allowed to settle in settling ponds in the form of a mixture of small phosphate particles and clay. This material is also sold, sometimes called soft rock phosphate or colloidal phosphate.

The hard rock phosphate takes some time for its nutrients to become available to plants. The soft rock phosphate, being colloidal, is a quickly available source of phosphorus, a small percentage at a time. But it also helps condition soil and prevent leaching of lime. It works well along with high-calcium lime, in a ratio of about 1/2 as much phosphate per acre as lime. A test in Georgia showed that rock phosphate can give forage yields equal to superphosphate, over a 5-year period.[7]

2. Next we have *biological products*, either living organisms or substances produced by them. Some of them not only improve soil structure, but also release plant growth stimulants.

a. *Microorganisms*. It is well known that many soil bacteria and algae produce sticky carbohydrates (polysaccharides) that glue soil particles together, giving good soil structure. A healthy, "living" soil should have plenty of microorganisms, but "dead" soil may have to be inoculated or stimulated. Some commercial products contain mixtures of bacteria species or bacteria plus fungi and yeasts. Some contain algae. Besides helping condition soil, certain microorganisms also fix nitrogen from the air, reducing fertilizer needs. Some commercial products contain microbial activators, mixtures of nutrients, enzymes, and other materials, intended to stimulate the growth of existing soil life.

Getting new species of microorganisms established in soil is not always successful, depending on soil conditions and substances toxic to the organisms. Also, sometimes species already present will out-compete or kill off the introduced species. Such products can give good results or none at all.[8] It has been found that just spraying about a quart of molasses per acre on soil or applying about one-half to one pound of granular

[6]G. H. Beames, Data Report no. AL-201 (undated), American Colloid Co., Skokie, IL.

[7]*Crops & Soils*, 1961, vol. 13, no. 5, p. 21.

[8]J. C. McAllister, *ibid.*

sugar per acre improves soil structure and stimulates soil life by serving as a food source for organisms.[9] No more than 3 gallons per acre per year of molasses should be used.[10] Adding any type of organic matter, if it is kept in the upper aerated layers of the soil, will stimulate microorganisms by serving as a food source.

b. *Organic matter*. Organic matter, either fresh or decomposed into compost or humus, does an excellent job of improving soil structure. Large amounts of fresh organic matter should not be incorporated into the soil or toxic by-products of microbial decomposition could be formed. Such materials as crop residues, animal manures, green manure, cannery wastes, sewage sludge, etc., are good sources. A mixture of animal and plant sources gives better results, and adding ground rock fertilizers is even better. Fresh organic matter should readily decompose to humus in the soil. The other alternative is to apply composts, which do not have the danger of toxicity. Besides improving soil structure, organic matter and compost increase soil life and supply many plant nutrients. Some companies specialize in compost technology and can sell compost, compost starters (bacterial inoculants), or compost-turning machinery. See Appendix B for details on composting.

Kelp or seaweed is one type of organic matter readily available near the oceans. If incorporated into the soil, it will not only condition soil, but also supply trace elements and growth-stimulating hormones (used at rates of 250–500 lbs./acre, vegetable yields generally increase from about 10 to over 100%).[11] Some companies sell seaweed meal or powder for farmers far from an ocean. Research at the University of Maryland found that seaweed meal on lawns greatly reduced nematode populations.[12]

c. *Enzymes*. Soil sprays containing microbial-produced enzymes have been found to improve soil structure, eliminate hardpans, correct extreme pH, increase the soil's nutrient-holding capacity (C.E.C., cation exchange capacity), and detoxify soil, as well as increase yields and quality.[13]

3. Finally, there are the *wetting agents*, sometimes called *surfactants*. A surfactant is a chemical that lowers the surface tension of water. A soap or detergent is a familiar example. There are a number of surfactant spray products for use on soil, which can increase the rate of soaking in of water, overcome a hardpan, and drain wet spots. Research has

[9] *Annual Review of Plant Physiology*, 1954, vol. 5, p. 78.
[10] *Acres U. S. A.*, September 1984, p. 23.
[11] W. A. Stephenson, *Seaweed in Agriculture and Horticulture*, 1968, p. 96.
[12] L. Fryer, *The Bio-Gardener's Bible*, 1982, p. 60.
[13] *Acres U. S. A.*, January 1985, p. 18–19.

shown that they work well on hydrophobic (water-repelling) soils, but not on hydrophilic (wettable) soils.[14] Generally, water-repellent soils are fine-textured. You can test this by placing a drop of water on the surface; if it beads up and soaks in slowly, a wetting agent may help your soil. Wetting agents should be used in moderation as a "band-aid" until you get more humus and microorganisms into the soil, because they can adversely change soil structure, reducing crumb structure and causing microorganisms to lose their protected homes inside soil clumps.[15]

Be different. Why grow the same crops as all your neighbors? Or raise the same animals? Probably the market price is too low to make a decent profit. Wouldn't you rather clear $1000 or more per acre? Or have a steady high income all year round? It's not hard to do. All it takes is some imagination and a willingness to be a little different from your neighbors.

If you grow high value or high demand items, you can receive premium prices or set your own price. Let your imagination run wild. Think of unusual or specialty items that will grow in your climate. Or find a new way to use an old crop. It is easier if you live in the South and near a large city, but greenhouses can be used in the North and marketing can be set up—or go into your own marketing business. Check on possible CO-OPs of other growers, or form a new CO-OP. Many people or companies cannot locate *high quality* food; they will beat a path to your door if you can provide it.

Some examples:

- Grow vegetables for supermarkets, restaurants, farmers markets, roadside stands, and health food stores; not just the ordinary varieties, but also unusual ones, perhaps to supply an ethnic market.
- Grow fruit, berries, or nuts; it can be done without excessive toxic spraying. Kiwi fruit (Chinese gooseberry) is becoming a popular crop in California.
- Grow flowers, tropical plants, seedlings, or seeds for nurseries, greenhouses, and seed companies.
- Grow herbs and seasonings.
- Grow forages for seed, either grasses or legumes.
- Grow unusual forages for livestock (Jerusalem artichoke, comfrey, carrots [for vitamin A], sunflowers [also produce seed and oil].
- In the corn belt, grow popcorn.
- Raise sorghum to make sweetener and syrup.
- Tap maple trees to make sugar and syrup.

[14]H. D. Sunderman, *Soil Wetting Agents*, North Central Regional Extension Publication 190, 1982.

[15]G. H. Beames, Research Bulletin, Nov. 1984, American Colloid Co., Skokie, IL.

- Grow trees for lumber, pulp wood, or fire wood.
- If you have a dairy, make high quality ice cream or yogurt.
- Raise goats for milk (needed by certain babies allergic to cow's milk).
- Raise fish for restaurants.

Horticulturist Booker T. Whatley of the Tuskegee Institute developed a plan for making small farms profitable. This can be done by planting about ten uncommon but marketable crops and selling them at an independent local market. By growing crops that mature throughout the year, a steady income will result. Some of Whatley's recommended crops include:

- For the Midwest: beans, asparagus, tomatoes, and highbush blueberry.
- For the Northeast: broccoli, cauliflower, Brussels sprouts, and spinach.
- For the West: English walnut, pistachio, hazelnut, and garlic.
- For the South: sweet potato, peas, blueberries, strawberries, mustard, collard and turnip greens, and black-eyed peas.

Crops should ripen at different times to avoid harvesting problems. Drip irrigation serves as a protection against drought and frost.

Whatley recommends a membership marketing system with about a thousand farms paying annual dues of $25 to protect against market fluctuations. He says that a 25 acre farm operated by two men can gross over $100,000 yearly within five years, since each crop should earn at least $3000 per acre.[16]

A questionnaire sent to 245 grocers, schools, hospitals, food service companies, and other bulk food buyers in two Pennsylvania counties revealed that they were willing to buy about $8.1 million worth of products from local farmers.[17]

New crops. Crop expert P. C. Mangelsdorf says only 15 crops feed the world: wheat, rice, corn, sorghum, barley, beans, peanut, soybean, sugarcane, sugarbeet, sweet potato, manioc, potato, banana, and coconut.[18] There are others, of course, but 90% of our food comes from a small number of species, and the number of crop varieties (genetic types) is rapidly decreasing. Some varieties are low in protein or essential amino acids. Pollution and destruction of natural habitats could result in the extermination of potentially valuable species and varieties.

A number of scientists and organizations are feverishly working to save seed from many less-used crop varieties and their wild relatives.

[16]*Acres U. S. A.*, September 1984, p. 4.
[17]*The New Farm*, May/June 1983, p. 14–16, 37.
[18]J. R. Harlan, *Crops and Man*, 1975, p. 80.

Also, there is increasing interest in developing a number of plants as new crops. Some have great potential to help overcome problems with increasing world population pressure.

1. *Winged bean*. The most promising new crop is the winged bean, a tropical climbing bean from Indonesia, New Guinea, and Africa. It is called winged because the pods have projecting flanges along the edges. It is truly a marvelous plant, sometimes called the "supermarket on a stalk." It germinates and grows rapidly, climbing to over 12 feet. It flowers in a month or less and pods can grow a foot long in two weeks, eventually measuring over two feet. It is tolerant of drought and most diseases and pests.

Almost every part of the plant is edible and high in protein, minerals, and vitamins. Pods are delicious and the seeds contain up to 42% protein (with nearly all essential amino acids) and 20% edible oils. Seeds can be made into flour, grits, hominy, tofu, cooking oil, milk and ice cream, lubricating oils, glycerin, varnishes, soap, etc. Leaves, flowers, and stems can be eaten, and tubers develop underground which can be eaten like potatoes, but which have 10 times the protein. Nutritious livestock and poultry feed can be made from all parts of the plant.

Although its potential for relieving world hunger is enormous, introduction of the winged bean into third world countries is lagging, sad to say. Indonesia and the Philippines are emphasizing it, and it has been introduced into over 70 countries, but governments often do not give it the priority it deserves. Jean-Pierre Hallet has introduced it to the Ituri Pygmies in Africa, with great success.

2. *Amaranth*. Grain amaranth was an established crop in the ancient Aztec culture. Native to Mexico and Arizona and a relative of the weeds pigweed and tumbleweed, drought-resistant grain amaranth gets up to 10 feet tall and bears large reddish or yellow seedheads weighing up to six or eight pounds and containing a half million seeds. The leaves taste like spinach and are rich in vitamins (A and C) and minerals (especially iron). The small seeds are protein-rich (over 15%) and high in lysine, an amino acid that is low in wheat and corn. Seeds can be ground into flour, sprouted, or popped by heating like popcorn. The Rodale Research Center, Kutztown, Pennsylvania, and the National Academy of Sciences are doing research in the U. S., while scientists in Peru, Guatemala, Mexico, Thailand, and Kenya are enthusiastic. Test plantings in Kansas yielded 230 pounds of grain per acre, and with no fertilizer or chemical input needed, grossed almost twice what wheat would bring. One grower in Peru received a 3600 lb./acre yield.[19]

[19] *The New Farm*, *ibid.*, p. 29–33.

3. *Jerusalem artichoke*. This member of the sunflower family is native to the U. S. Growing 6 to 12 feet high, the plant's coarse leaves and tops make nutritious livestock feed. The underground tubers can be eaten like potatoes or be used for livestock. It is very hardy and spreads by sprouting roots. Despite some marketing problems, the Jerusalem artichoke has been increasingly grown in the U. S. in recent years.

4. Other less well known food crops include the *tepary bean*, native to Mexico and Arizona, and producing high-protein beans; the *groundnut*, a viny legume growing near lakes and rivers in the eastern U. S. and bearing lemon-sized tubers containing three times the protein of potatoes; *buffalo gourd*, a desert vine suitable for forage, which has seeds with 35% oil plus high protein, and roots with 20–50% starch; *ragi* (finger millet), a nutritious (similar to wheat or sorghum) small grain grown in Africa and India and adaptable to the southern U. S.; *cocoyam*, a tropical plant that can be baked, fried, or ground into flour; *pigeon pea*, a high-calorie crop with 20% protein; and *Chinese milkvetches*, legumes valuable for livestock fodder, green manure, and controlling wind and water erosion. They have been introduced into eastern Washington by agronomist Dr. Tom Lumpkin. Used in rotation with corn, corn yields were increased while soil was improved. Lumpkin said that the use of chemical fertilizers alone does nothing to improve the productivity of the soil because they do nothing to hold the topsoil in place.[20]

5. Some new crops are not useful as food, but have other values. These include *jojoba*, a desert shrub native to southern California, Arizona, and Mexico. It bears seeds containing 44–58% of a high grade oil ("liquid wax") that can be used for lubricating motors, waxing floors, cooling transformers, moisturizing skin, making candles, and feeding livestock. The oil does not become rancid and may serve as a vegetable oil in human foods. It is often mentioned as a substitute for sperm whale oil.

Guayule is another desert shrub from southwestern Texas and northern Mexico, whose sap can be used to make rubber. It was planted in the early 1900's and during World War II, and is again being grown for this purpose.

Euphorbia, a relative of the household poinsettia, produces hydrocarbons in its sap which can be made into fuel and plastics. A fast-growing variety of *sugarcane* is being developed to use as a fuel for boilers in electric power plants.[21]

Open pollinated crops. We have previously covered the problems with hybrid crop varieties and the superior quality of open pollinated varieties in Chapters 1 and 5. Hybrids do not "breed true;" new seed

[20]*Science*, 1984, vol. 225, p. 1445–1448; *Farm Journal*, February 1982, p. B-2; R. Wolf (ed.), *Organic Farming*, 1977, p. 127–128; *Farmer-Stockman of the Midwest*, October 1, 1984, p. 8; *National Geographic*, May 1981, p. 702–712.

[21]J. W. Rosenblum (ed.), *Agriculture in the Twenty-first Century*, 1983, p. 146–149; *Science*, 1983, vol. 219, p. 24–26; 1984, vol. 225, p. 1445–1448; *Farm Journal*, February 1982, p. B-2; *National Geographic*, May 1981, p. 702–712.

must be purchased every year because not all offspring from planting hybrid seed are like their parents. Open pollinated seed can be saved and planted year after year. Hybrids are bred to give high yields, but are to some degree unbalanced in their food value. Open pollinated varieties can yield nearly as well as hybrids in good soil, and they are of superior quality as food.

Traditional methods of improving open pollinated varieties are slow and uncertain. They involve selecting seed from the best-looking plants (see H. Willis, *How to Grow Top Quality Corn*, 1984, Chapter 6). Recently, methods of growing open pollinated crops in fields subjected to weak electromagnetic energy (see later in this chapter) have produced plants and grain with greatly improved nutritional value, with regard to digestibility, amino acid balance, and enzyme content.

Better crop growth. There are a number of exciting methods that are being developed (others are older but increasing in application) which increase the growth, health, yield, and quality of crops. They include seed treatments, growth stimulants, foliar feeding, microbial "helpers," natural pest and weed control, efficient cropping systems, effective fertilizing materials, and the use of weak natural energies.

1. *Seed treatments.* There are various kinds of seed treatments with different purposes. Some are coatings of fungicides or insecticides to prevent soil-borne diseases or pests from attacking. Although such poisons should not be needed in healthy, fertile soil, they can be effective in less than ideal soil and under unexpected stress conditions (waterlogging, cool weather); also, smaller amounts of poisons are used than if all the soil were treated.

Inoculation of seeds with beneficial microorganisms has been done for years in the form of root nodule bacteria (*Rhizobium*) for legumes, with excellent success. Inoculation with general beneficial root bacteria (rhizosphere bacteria) and fungi (mycorrhizae and nematode-trapping fungi) is still somewhat experimental, but is being developed commercially and will definitely be important in the future. These microorganisms can help roots fight harmful microbes and nematodes, and help roots absorb soil nutrients (see Chapter 2).

Then there are commercial seed treatments that contain growth-stimulating enzymes, hormones, and trace elements. These get the germinating plant off to a faster, more vigorous start and stimulate a bigger root system, which are important in cool or northern climates. Tests at Clemson University found that soaking beet seeds in seaweed extract for 30 minutes increased germination 25%.[22]

[22]W. A. Stephenson, *ibid.*, p. 106–107.

2. *Growth stimulants and regulators*. Plant growth regulators are either natural or synthetic substances that act in small amounts to affect plant functions so that growth is increased (or sometimes flowering or fruit drop is delayed), quality is improved, yield is increased, or harvest is facilitated. Natural growth regulators include plant hormones and seaweed extracts. They often result in increased yield, quality, disease and pest resistance, and cold hardiness[23], but if the soil and plants are already healthy, there may be little effect.[24]

There are also some products called activated water or catalyst altered water, made by treating water with electricity or an extract of lignite coal. The molecular configuration of the water is said to be changed, making it more active. When such treated water (diluted with ordinary water) is applied to soil, seeds, or foliage, plant growth is stimulated. It can also be added to livestock feed and used for human health.[25]

Some synthetic growth regulators act as herbicides for some species and growth stimulants for others. Too much of a growth regulator or applying it at the wrong time of plant development or in the wrong weather can cause growth inhibition or plant damage. Some products increase yields of some crops, but not others. Synthetic chemicals may be toxic to fish or wildlife.[26]

3. *Foliar feeding*. We have previously covered (Chapter 5) foliar feeding to supply small amounts of nutrients, mainly trace elements. Also, we mentioned that a crop can be induced to "switch over" from vegetative growth to begin flowering and seed production by the right foliar spray. Another use of foliar spraying is to apply some of the growth regulators covered in the previous section. Since they are needed in only small amounts and at certain times, foliar spraying is the best method of application. Such sprays are often used on fruit trees to increase blossom set, and to improve shipping and keeping quality, fruit color, and ripening.

4. *Microbial helpers*. Besides the use of microbial inoculants as seed treatments, covered in section 1, similar inoculants are being developed for general soil application. Nitrogen-fixing bacteria and algae which do not have to live in root nodules are one approach; they can benefit non-legume crops. A general population of beneficial soil bacteria or the special rhizosphere species is another approach. A third is the special symbiotic fungi, mycorrhizae. If these beneficial microorganisms become successfully established in the soil, they should remain active for years

[23] *Ibid.*, p. 95–124.
[24] *The New Farm*, July/August 1980, p. 29–35.
[25] *Acres U. S. A.*, November 1984, p. 14–16.
[26] J. C. McAllister, *ibid.*, p. 85–98.

(no toxic chemicals should be added, of course). It is often difficult to get the microbes successfully established, however, because microbes already in the soil "out-compete" them. An excellent inoculant of beneficial microbes is compost; another is poultry manure. Still another is using water that has had animal manure or plant residues (old hay, weeds, or grass clippings, etc.) soaking in it. The special algae mentioned under soil conditioners earlier in this chapter also can improve growth and yield.

5. *Natural pest, disease, and weed control.* We have previously mentioned allelopathy (Chapter 4), the ability of the secretions or residues of one plant species to inhibit or kill another. Some weeds are allelopathic, quackgrass for example, and can seriously stunt crops. Certain crops or their residues are known to inhibit other plants, including corn, sorghum, oats, barley, rye, tomato, potato, peas, beans, cucumber, asparagus, apple, peach, apricot, and citrus. The proper management of these "natural herbicides" can reduce or eliminate the need for ordinary herbicides.[27]

And of course, the old practices of controlling weeds by timely cultivation or frequent mowing are very effective. Cultivation has the added benefit of breaking a crust and aerating the soil.

Healthy plants have built-in mechanisms to resist diseases and pests (see Chapter 4). We have mentioned how root-inhabiting bacteria and fungi protect against soil-borne diseases. Recent research has shown that plants produce various chemicals that repel or kill invading pathogens or pests. Some of these protective chemicals are only produced when microbes start to invade or pests attack.[28] Insects raised on plants that had previously had their leaves removed or had suffered pest attack ate less and had reduced survival rates.[29] Then there is the amazing phenomenon of "whispering trees," the discovery that when insects eat the leaves of one tree in a forest, neighboring trees produce bitter tasting tannins in their leaves, because of an air-borne "chemical messenger."[30]

The concept of integrated pest management (IPM), mentioned in Chapter 6, is the right approach toward short-term pest control problems (the best long-term approach is building up healthy soil so plants will be healthy and naturally resist pests, and encouraging natural enemies and beneficial soil organisms). In integrated pest management, the idea is not to completely eradicate a pest species, but to keep its popula-

[27] J. W. Rosenblum, *ibid.*, p. 347–348.

[28] B. J. Deverall, *Defense Mechanisms of Plants*, 1977; M. Kogan & J. Paxton, *Plant Resistance to Insects*, 1983, p. 153–171.

[29] *Science*, 1984, vol. 225, p. 53–54.

[30] *Science*, 1983, vol. 221, p. 277–279; G. H. Orians & D. F. Rhoades, *National Science Foundation News*, 1983, PR83–41, 2 pp.

tion levels below the point of doing serious damage. This is done by using non-toxic controls when possible (natural enemies, management practices) and toxic chemicals only when necessary.

6. *Efficient cropping systems*. Where the growing season is long enough, two, three, or more crops can be grown yearly (called multiple cropping). Of course, the soil must be fertile enough to supply the necessary nutrients, and there must be enough soil moisture (if rainfall can't supply it, irrigation may be needed). Soil will warm up faster in the spring if it has a high humus level (and high microorganism activity). A light application (200 lbs./acre broadcast) of ammonium sulfate in the fall will also do the same thing. The most efficient multiple cropping systems choose certain crops and time their growth so that they make optimal use of available sunshine and temperatures (that is, grow cool-season crops such as wheat in spring or fall and heat-loving ones such as soybeans in the summer).[31] Productivity per acre can be increased as much as 30% in this way.[32]

Mixing two or more crops in the same field is called intercropping. Planting species that benefit each other (or one benefiting the other) is called companion planting, a popular thing with gardeners. Soybeans (or some other legume) and corn are a good combination. The legume's root nodules enrich the soil with nitrogen (if soil conditions are not toxic for them). If corn rows are wider than usual, with soybeans between them, the corn yields per acre are about the same as corn alone, but the shading from the corn stunts the soybeans slightly, giving a reduced yield (but still, you get two crops off the field, not just one).[33] Another system is to alternate two or four rows of corn (with narrow spacing) with four or six rows of soybeans. In this case, the corn yield is much higher because most rows are edge rows and receive more light, and the soybeans do better as well (this system is popular in China and Taiwan, where good land is scarce).[34]

Planting a forage such as alfalfa along with a small grain or row crop is a common practice. The young forage plants get a head start and protect soil from erosion.

Intercropping different plant species or even different varieties of the same species can reduce insect pest attack. Tomatoes planted along with Brussels sprouts reduced the number of eggs laid by two insect pests of Brussels sprouts.[35]

[31]*Successful Farming*, May 1982, p. 16–17; *Farm Journal*, March 1982, p. 38.
[32]J. W. Rosenblum, *ibid.*, p. 141.
[33]R. Wolf, *ibid.*, p. 115–117.
[34]J. W. Rosenblum, *ibid.*
[35]R. M. Perrin & M. L. Phillips, *Entomol. Exper. et Applic.*, 1978, p. 585–593.

7. *Effective fertilizers*. As an alternative to using large amounts of synthetic commercial fertilizers, which are often expensive or may harm soil organisms or soil structure, there are several fertilizer materials and methods to consider. Some of the soil conditioners we covered previously in this chapter also have fertilizer value. These include liming materials, rock phosphates, other ground rock trace element fertilizers, humates, seaweed, compost, and other organic matter. Applications of finely ground glacial gravel (or gravel crusher screenings) provides a "shot in the arm" of major and trace elements, giving terrific crops: two to four times increase in yield of garden produce in California (including 18 foot pole beans), 12 foot clover in Vermont, and a 15 bushel/acre increase of corn in Michigan.[36] You can make an effective liquid fertilizer for nothing by mixing organic matter (such as manure, old hay, garbage) in a stagnant pond or water tank and waiting until the "brew" is good and ripe. The decomposing organic matter will leach nutrients into the water, and if nitrogen-fixing bacteria are active, you can get a great increase in nitrogen content. Dilute and strain the mixture before spraying on your land.[37] Adding some compost to the mixture would be an excellent way to introduce beneficial microorganisms.

Careful use of commercial "N-P-K" fertilizers can give good results, provided the right materials are chosen. Materials that harm soil structure (anhydrous ammonia) or harm soil organisms or seedlings (anhydrous ammonia, chloride-containing materials, high salt materials) should be avoided. Desirable sources of nitrogen include ammonium nitrate, ammonium sulfate, ammonium thiosulfate (used in liquids), calcium nitrate, potassium nitrate (from Chile, not the American brand), and mono-ammonium phosphate. Sources of phosphorus are superphosphate (not triple superphosphate), mono-ammonium phosphate, and in liquids, ammonium polyphosphate and phosphoric acid. Potassium is best obtained from potassium sulfate or Chilean potassium nitrate. Which materials should be used and the amount depends on many factors: the crop, time of year, past fertilizing history, method of fertilizing (broadcast or row), and soil needs shown by soil tests. Generally, if possible, commercial N-P-K materials should be used in smaller amounts to *supplement* natural fertilizers or for special uses, such as side dressing in mid-season to supply needed soil "energy" (ionization) or to "switch" a crop from vegetative growth to seed production (covered in Chapter 5).

If the soil is deficient in one or more trace elements, application of

[36]J. D. Hamaker, *The Survival of Civilization*, 1982, p. 40–43.
[37]*Crops & Soils*, March 1977, p. 23.

the right materials can make a big difference in yield and quality. The potential for yields in really good soil is amazing. On test plots and fields in Idaho and Washington, open pollinated corn has yielded over 400 bushels per acre, and dry-land wheat over 250 bushels per acre (with 19.6% protein).[38]

8. *Natural energies*. There is a whole world of energy surrounding us and influencing plant, animal, and human growth and well-being. We literally live in a sea of energy. Besides the obvious light energy we receive from the sun, there are also invisible ultraviolet and infrared light wavelengths, and natural and man-made radio and microwave, X-rays, and gamma rays. These are called electromagnetic radiation, and they occur in different wavelengths or frequencies (see diagram). Everything

THE ELECTROMAGNETIC SPECTRUM

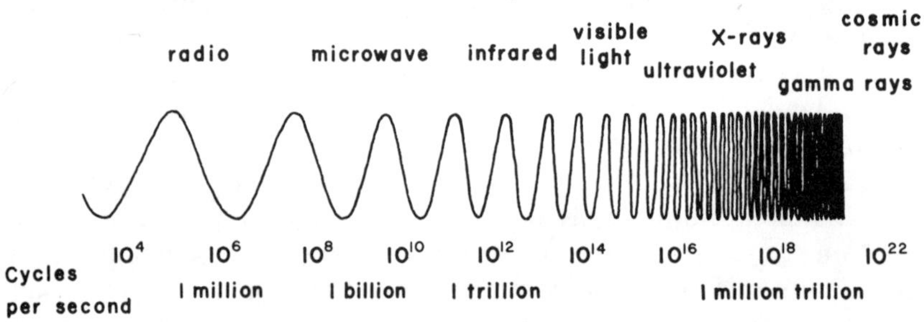

in the universe—stars, rocks, plants, and ourselves—radiates electromagnetic waves at a frequency depending on its temperature, which is due to the vibrations of the atoms or molecules of which the object is composed. Very hot objects radiate infrared, visible light, ultraviolet, X-rays, or gamma rays. Cooler objects give off infrared, microwaves, or radio waves.

Besides these energies, there is electrical energy: static electricity, lightning, d.c., and a.c. Electrical charges possess a force field around them. Fluctuating electric charges can be used to generate electromagnetic radiation; this is how a radio station broadcasts.

The earth has a magnetic field. All objects on the earth are subjected to it. A change or movement of an electrical charge generates a magnetic field and a change in a magnetic field generates an electrical field. Charged particles (electrons, protons, ions) in motion are affected by a magnetic field, and charged particles themselves can create an electrical

[38] *Acres U. S. A.*, April 1984, p. 32; talk by Bill Johnson, Soil Spray Aid, Inc., Moses Lake, Washington, June 1983.

field that affects living things. It is now known that an excess of positive ions (common in modern buildings and certain dry south winds) at first causes people to feel energetic, but after a while they "overload" the body's metabolic processes and hormone system, leading to anxiety, migraine headaches, arthritis flare-up, asthma, high blood pressure, higher accident and suicide rates, etc. Negative ions (common in mountains, forests, and near waterfalls) or a "natural" balance of five positive to four negative ions, cause a sense of calmness and tranquility, sound sleep, and help heal burns, asthma, etc. Plants grow unusually large with positive ions.[39]

Cosmic rays (nuclei of atoms stripped of their electrons) bombard the earth from outer space, along with other charged particles, protrons and electrons from the sun (the solar wind). The beautiful aurora borealis (northern lights) result when these particles interact with the earth's magnetic field and atmosphere. The sun also has a magnetic field which is variable and reaches and influences the earth's field.[40]

Living cells and the tissues and organs they compose generate weak electrical and magnetic fields. Ions (electrically charged molecules) can be held or "pumped" to one side of a cell membrane, creating an electrical charge. This is the origin of the electrical impulses in animal and human muscle and nerve cells.[41] Plant cells also generate electrical currents. The inside of pea root cells is about - 110 millivolts compared to the outside, while oat root cells are about - 84 millivolts.[42] The microscopic threads of a fungus produce currents that enter at the tips and exit farther back.[43] Even the various membranes within cells are charged because of unequal sodium, potassium, and chloride ions and act as microscopic capacitors.[44] Evidence indicates that the self-generated electric fields of a growing plant help control its development by distributing growth hormones.[45]

Weak electrical currents and electromagnetic fields are known to affect cell activity. The right kind of current can speed up cell functions.[46] Microwaves have been found to be absorbed by DNA, the material that makes up genes, with possible damage and cancer initiation.[47] Weak

[39]F. Soyka, *The Ion Effect*, 1977.
[40]*Science*, 1984, vol. 225, p. 793–800.
[41]H. Breithaupt, p. 13, in *Electromagnetic Bio-Information*, 1979.
[42]F. B. Salisbury & C. Ross, *Plant Physiology*, 1969, p. 155.
[43]*Science*, 1983, vol. 220, p. 1385–1387.
[44]H. Breithaupt, *ibid.*
[45]*Scientific American*, October 1962, p. 107-117; E. J. Lund, *Bioelectric Fields and Growth*, 1947, p. 281–289.
[46]*Science*, 1983, vol. 220, p. 1283–1285.
[47]*Science News*, 1984, vol. 125, p. 248; AP news release, August 17, 1984.

electromagnetic fields can alter the flow of calcium ions in brain cells, cause animals' body temperatures to rise, increase the number of white blood cells, and slow down learning. Scientists are concerned about adverse effects on people living near radio broadcasting towers, large power lines, and other sources of extremely low frequency (ELF) radiation.[48]

Sending low-voltage pulsating electrical currents into the body through electrodes on the skin has been able to control the pain of neuralgia, headache, childbirth, and arthritis, as well as helping circulatory problems, speeding healing of bones and burns, and improving memory.[49]

Way back in the early years of this century, experimenters in England treated seeds with weak electric current for several hours (seeds were soaked in a weak brine to carry electricity). Results included an increase of 6 to 19 bushels/acre of oats, 6 to 12 bushels/acre of wheat, and 21 to 50% increase of barley. More recent experiments along the same line have been done, with promising results.[50]

Recently, iron particles (magnetite) have been found in the cells of certain kinds of bacteria and algae; and in the heads of fish, sea turtles, birds, rodents, monkeys, and humans. This may be how birds and water creatures orient themselves during migration.[51]

Living things are affected by external magnetic fields. Seeds placed in a moderate magnetic field for 10 days germinate faster than untreated seeds, and the seedlings grow faster (for strong fields, a shorter exposure time is sufficient). For example, in four days, magnetically treated barley roots grew 34.72 mm long and the shoots 21.10 mm, compared to 29.35 mm and 15.51 mm for untreated plants. Treated soybean roots grew 22 mm and the shoots 88 mm, compared to 6 mm and 68 mm for untreated plants. Results are better if the long axis of the seed is parallel to magnetic lines of force. Some crops mature earlier and produce higher yields when magnetically treated. The cells of magnetically treated plants apparently function more efficiently, producing more sugar but using less energy to metabolize. More dividing cells are present in the growing tips of roots and stems. Seedling roots grow out parallel to magnetic lines of force and grow away from a stronger magnetic field and toward a weaker one.[52] Canadian research found that some plants' roots will mainly grow

[48]*Science*, 1984, vol. 225, p. 32–33.

[49]*Changing Times*, January 1982, p. 50–54.

[50]C. Mercier, *A Manual of the Electro-Chemical Treatment of Seeds*, 1919; *Acres U. S. A.*, 1976 Fair Issue, p. 4; September 1981, p. 4.

[51]*Science*, 1982, vol. 215, p. 1492–1493.

[52]*The Science Teacher*, May 1973, p. 57–60; *Economic Botany*, 1968, vol. 122, p. 124–134; *Soviet Plant Physiology*, 1965, vol. 12, p. 904–911; 1975, vol. 22, p. 84–89.

outward in a north-south plane (winter wheats, some spring wheats, wild oats, some weeds), while others do not (barley, flax, corn, sunflowers). If crops whose roots do grow north-south are planted in rows running east-west, they will more effectively use between-the-row soil nutrients and water.[53]

Some researchers claim that the south pole of a magnet increases growth and that the north pole decreases it (the south pole is defined as the pole that turns toward the earth's north pole, since opposite poles attract). In one experiment, crops grown from seed treated with the south pole of a magnet had better growth, yield, protein, sugars, and oils than those treated with the north pole. Laboratory rats living in south-pole magnetism were larger and healthier than those living in north-pole magnetism, although they died sooner.[54] Some people claim to have been cured of various ailments by wearing small magnets next to their bodies, including arthritis and cancer.[55] Weak magnetic fields, such as the earth's magnetism, are known to influence cell growth, life span, and activity of organisms.[56]

Different molecules or substances have a "natural frequency," and they absorb electromagnetic waves much more readily at that frequency than others.[57] A number of people speak of cells or whole plant or animal bodies as having a certain frequency. In the 1920's the Russian engineer, Georges Lakhovsky theorized that the nucleus of a cell is like an electrical oscillating circuit, emitting and receiving electromagnetic vibrations at a certain frequency if the cell was healthy. If the cell was unhealthy or dying, there was a lower frequency. He built machines emitting short wavelength energy and was able to cure plant cancers, make plants and animals grow faster, and cure hospitalized patients of cancer, radium burns, goiter, and other "incurable" ailments. The best results were obtained with a machine called a multiple wave oscillator, which allowed every cell in the body to vibrate at its own best frequency.[58]

Towers, etc. In recent years, Jerry W. Fridenstine has invented a device called an orthomolecular multiwave oscillating generator (often called a tower), which is said to "broadcast" certain frequencies (the individual frequencies of each crop and animal species) over a farmer's land. Some impressive results have been obtained, such as alfalfa growing 48 to 56 inches high and yielding over five tons per acre on the first

[53]*Crops & Soils*, 1968, vol. 20, no. 8, p. 8–10.
[54]A. R. Davis & W. C. Rawls, Jr., *Magnetism and Its Effects on the Living System*, 1974, p. 31, 37–39.
[55]U. Warnke & F. A. Popp, p. 197, in *Electromagnetic Bio-Information*, 1979.
[56]*Ibid.*, p. 196.
[57]*Science News*, 1984, vol. 125, p. 248.
[58]P. Tompkins & C. Bird, *The Secret Life of Plants*, 1973, p. 199–201.

One of Jerry Fridenstine's towers. Parts of the tubing are filled with special mixtures of minerals.

cutting, full silos on much less acreage, over 200 bushel per acre corn in Colorado, more digestible animal feed, and contented, healthy dairy cattle, with an increase in milk production. The towers are said to focus and broadcast natural energy in the earth's surface, to sort-of "perform acupuncture on the earth" in order to balance out too high or too low elements in the soil or animal feed.[59] A somewhat different device has been invented by Dr. T. Galen Hieronymus, consisting of a vertical pipe with the bottom buried in the ground. A slanted collector plate at the top is said to collect "cosmic" or "eloptic" energy and transmit it into the soil through a radiating coil at the bottom. It causes plants to grow faster, be healthier, and produce larger, higher quality yields within a radius of

[59]*Acres U. S. A.*, June 1984, p. 30–32.

about 1/2 to over one mile. Results include five-foot high alfalfa (double the previous year's yield in one cutting) and five-foot marigolds.[60] Fridenstine and others say that electromagnetic energy flows from the equator to the poles through the organic-matter-rich soil. This energy helps plants to grow better. Some research indicates that there are north-south lines of greater energy (called ley lines), and that trees and other vegetation grow larger along these lines. The energy also goes in or comes out of the earth's surface at certain places, called vortexes. Man's activities and structures, such as roads, pipelines and power lines, interrupt the normal flow of "earth energy." Increasing the humus content of your soil will increase the energy available to plants. The commercial salt fertilizers of modern agriculture make plants grow fairly well because salt is a good electrical conductor, increasing energy flow. But at a price. Soil bacteria and fungi are disturbed or killed, humus is depleted, and the soil becomes "dead," sterilized.

The same type of thing can be done on a small scale, such as in your garden or flower beds. There are various ways of concentrating or focusing the "earth energy." In 1925, Georges Lakhovsky set up a special coil of copper wire (one foot in diameter) around a geranium plant. The plant grew—and grew—and grew. By 1928, it was four and one-half feet tall.[61] Another method is to stick two 18 inch copper wires or rods or large nails into the soil north and south of a plant. Or simply sprinkle iron filings on the soil. Growing plants on raised beds is supposed to increase energy.

The source of this plant-growing energy is the sun. Besides the light and other forms of energy emitted by the sun, the weak electromagnetic energy of the soil streams out from the sun. Dr. Philip Callahan, a researcher well known for his work on how insect pests locate the plants they attack (see his book, *Tuning in to Nature*, 1975), has recently expounded a theory on how it works. In his book, *Ancient Mysteries, Modern Visions* (1984, p. 104–109), Dr. Callahan says the energy from the sun is a combination of magnetic and electrical charges called magneto-electric monopoles. A monopole is like one end of a magnet. There are two kinds, a north magnetic pole combined with a positive electrical charge, and a south pole with a negative charge. Both kinds stream from the sun and strike the earth. The north monopoles are mainly absorbed by plants and move down into their roots. The south monopoles are absorbed by many kinds of rocks and soil, and they can flow along in the soil. When the south monopoles in the soil meet the north monopoles in

[60]*Acres U. S. A.*, December 1984, p. 23–26.
[61]P. Tompkins & C. Bird, *ibid.*, p. 200–201.

roots or a seed, energy is released that causes plant growth. Nitrogen acts as a catalyst to help the process occur.

The more fertile soils are highly paramagnetic and attract more south monopoles (a paramagnetic substance is a non-metallic substance that is weakly attracted by a strong magnet). Clay and volcanic soil are highly paramagnetic and very fertile. Plants grow better in clay flower pots than plastic ones. Many of the ancient stone towers and pyramids around the world are excellent collectors ("antennas") of paramagnetic energy. Plants grow better near them than elsewhere.

A related idea is "pyramid power," the energy that is apparently focused on the inside and released from the apex and edges of a four-sided pyramid with the same proportions as the pyramid of Cheops in Egypt. Pyramid expert Les Brown (see his book, *The Pyramid*, 1978) states that magnetic energy, cosmic rays, and radio waves are focused by a pyramid. He has grown 12 or 13 pound cabbages; 21 inch, four pound cucumbers; and 50–60 pounds of tomatoes per plant under a pyramid-shaped greenhouse. The same types of plants grown outside produced three pound cabbages; 14 inch, one pound cucumbers; and 10–14 pounds of tomatoes per plant. There were no pest problems inside the pyramid. Food placed under a pyramid will gradually dehydrate, but will not rot.

A sound idea. Still another exciting use of natural energy to stimulate plant growth is the use of certain types of sound. Sound waves are a mechanical movement of air molecules or other matter, and are not electromagnetic in nature. Nevertheless, sound can have an effect on living organisms. It is well known that certain types of music (especially rock-and-roll) have harmful effects on our bodies, including our heartbeat, brain waves, and glandular secretions, while most "classical" music has beneficial effects. The wise farmer will tune the radio in his dairy barn to soothing music.

Sound also affects plant growth. A number of early experiments, as well as ancient Hindu lore have prompted modern investigations. In 1960, agricultural researcher George E. Smith planted corn and soybeans in two greenhouses. In one he played a recording of "Rhapsody in Blue" continuously. The serenaded plants germinated sooner than those in silence. They also had thicker, greener stems, and ten of the corn plants produced 40 grams of top growth, while the silent-treatment plants only produced 28 grams. The following year a field test of a hybrid corn variety resulted in 137 bushels per acre for musically stimulated corn, compared to 117 for non-stimulated corn.[62]

In the late 1960's, professional musician Dorothy Retallack did

[62] *Ibid.*, p. 161–165.

carefully controlled experiments (for a college biology course at first) and found that plants will literally wither and die when forced to "listen" to rock-and-roll music (what effect, then, does it have on us?), while classical music, jazz, and sitar music from India stimulated plants.[63] Just a continuous sound tone of 3000 cycles per second (a low hum) increased growth rates of a variety of plants and caused some house plants to bloom six months early, according to Dr. George Milstein, horticulture teacher at the New York Botanical Garden.[64] Canadian researchers tested early growth (up to eight weeks) of two varieties of wheat at two sound frequencies, 5000 and 12,000 cycles per second. Plants grew much more vigorously when exposed to the sound, especially one of the varieties (Rideau) at 5000 c.p.s. The weight of both roots and tops was 250–300% greater than controls.[65] They hypothesized that such low levels of energy could cause functional effects in cells if cell parts vibrate at a certain natural frequency. Sound at that frequency would cause much greater vibration (resonance) and add much more energy to the cell.

More recently, Dan Carlson, a Minnesota plant breeder, has developed a method of combining sound with a special foliar spray to cause fantastic plant growth. Starting with a purple passion house plant which grew 1400 feet long in 2 1/2 years (they normally only grow 18 inches tall), Carlson went on to such feats as a 15-foot tomato plant bearing 836 tomatoes; double yield of potatoes; 100–200% yield increase of sweet corn in 30 days, with the corn being much sweeter and better tasting; and 300 or more pods per soybean plant, with 7–10 pods per cluster. Treated seed germinates faster and produces a larger root system. Crops mature earlier; in field tests, potatoes in Minnesota matured by July 1, and grapes matured two weeks early. Cabbages had no insect damage.

For home or garden use, Carlson has cassette tapes of his oscillating high frequency sound covered up with Indian sitar or classical music. For agricultural use, he has a sound emitter that attaches to the back of a tractor that can cover about 35 acres (standing still) with its piercing, canary-like sound. About 15 minutes after the sound is emitted, the crop is sprayed with his special six-ingredient natural foliar spray. For field crops, three sprayings about two weeks apart are effective. Trees may require more. Tests with radioactive tracers have found that plants absorbed over 700% more spray than controls.

Carlson believes the sound opens up the plants' leaf pores (stomata) so they can "breathe better" and absorb more spray. The sound alone or

[63]*Ibid.*, p. 168–172.
[64]*Ibid.*, p. 173–174.
[65]*Canadian Jour. of Botany*, 1968, vol. 46, p. 1151–1158.

the spray alone give good results, but together, they produce amazing results.[66]

The moon's effect. You have all heard of the old custom of planting seed according to the phases of the moon to give better crops. Modern science has pooh-poohed such "ignorance." But recent research has revealed there is something to it. One researcher found that the moon does indeed affect the physiology of growing seedlings.[67] Much experimenting by M. Thun and H. Heinze in Germany in the 1960's and 1970's discovered that yield is related not to the four quarter phases of the moon, but to the lunar sidereal month of 27.3 days (the movement of the moon in a 360° circle in relation to the positions of fixed stars). They found that seed crops do best if planted when the moon is at 30°, 150°, or 270° apparent longitude. Leaf crops do best if planted when the moon is at 0°, 120°, or 240°. Root crops do best if planted when the moon is at 60°, 180°, and 300°. The longitude of the moon for each day of the year can be found in an annual "Ephemeris" or set of astronomical tables, available in many libraries. However, days when there are eclipses, apogee (when the moon is farthest from the earth), perigee (when the moon is nearest the earth), or other unusual events should be avoided.[68]

We all know the moon affects the tides by gravitational action. It apparently also causes the earth's natural magnetic energy to fluctuate, which may be partly why plants grow better or worse at certain times.

The fringe. Another use of weak energies in agriculture (and health) is the controversial practices of *radionics* and *psychotronics*. Since each type of atom and molecule radiates a different frequency of energy, it is believed that each plant, animal, or person (or their organs) radiate an energy field that is the sum total of their individual atoms. Inanimate materials also radiate at characteristic frequencies. If a plant, animal, or human (or certain cells or organs) is sick, the energy field is altered, the "vitality" is lowered. In the early part of this century, a detector machine was developed by Dr. Albert Abrams which apparently could diagnose human diseases, even from a single drop of blood.

Abrams then reasoned that if sick tissue radiated at a different frequency from healthy, then one should be able to "broadcast" the healthy frequency and help the body to cure itself. Or if there was an infection of germs, broadcasting the frequency of a material that will kill them (such as quinine for malaria) would eradicate the disease. He built

[66]*Acres U. S. A.*, November 1984, p. 1, 6–7; talk by D. Carlson, November 28, 1984.

[67]P. Tompkins & C. Bird, *ibid.*, p. 211.

[68]H. H. Koepf, B. D. Pettersson, & W. Schaumann, *Bio-Dynamic Agriculture*, 1976, p. 204–206.

such a machine and it was apparently successful, so much so that the medical establishment rose up to denounce him.[69]

More recent applications of such machines (called radionics machines) have reportedly been able to achieve up to 80 or 90% control of insect pests in fields by broadcasting the frequency of an insecticide, but not actually using any on the fields (neighboring fields or test plots were decimated by the insects). Even more amazing is that the machine is said to be able to do this at a distance by "treating" an aerial photograph of the field. Radionics operators claim that the emulsion of the photograph captures the frequency of the growing crop. When the USDA investigated such goings on, they and pesticide manufacturers teamed to call radionics a fake, causing the company making the machines to go out of business in the early 1950's.[70] Diseases can also be diagnosed from a photograph of the patient.

During experimentation in the 1950's, it was found that some people are able to cause plants to grow better without even using a radionics machine—just by mental power, mind over matter. This has developed into what is called psychotronics, the influencing of organisms, with or without the aid of a radionics machine, but guided by the operator's mind. Dr. T. Galen Hieronymus and Jerry Fridenstine have been among the leaders in using radionics and psychotronics for agriculture.

Because of the possibilities for the use of such methods for evil as well as good (which is admitted by the methods' users), I question the desirability or even the necessity of their use. There are plenty of other effective methods that rely on natural laws and energies alone. In the wrong hands, psychotronics could be used to cause pests and diseases to spread and crops to fail. It could replace germ warfare. It is nothing to play around with.

Alchemy? Can a chicken make calcium from potassium? Can plants and soil bacteria manufacture calcium from silicon, which is part of sand? If such changes in elements seem to defy the laws of chemistry and suggest medieval alchemistry, don't forget that naturally occurring radium disintegrates into lead and helium. Using high energy atom smashers, nuclear physicists can change one element into another and even create new elements. For example, bombarding nitrogen with helium nuclei (alpha particles) produces oxygen and hydrogen.

Experiments by a number of scientists over the last hundred years indicate that living organisms can change or transmutate elements using

[69]P. Tompkins & C. Bird, *ibid.*, p. 328–333.
[70]*Ibid.*, p. 333–343.

some mysterious low-level energy that is different from the high energy that nuclear physicists and atomic bombs use.

Most biologists and biochemists believe that life is essentially chemistry, that living cells are just miniature machines running on chemical energy. Ordinary chemical changes involve the electrons spinning around the central nucleus of an atom. Two atoms of hydrogen, for example, combine with one atom of oxygen to form a molecule of water by sharing electrons. In other cases, one or more electrons are transferred from one atom to another (forming ions).

But although scientists can "take apart" a cell and find out what elements it is made of, they cannot create life. Life seems to contain an extra "spark" of energy—it is more than the sum of its parts.

French scientist Louis Kervran fed a chicken only oats. He measured the amount of calcium in its eggs and droppings and found that the chicken produced four times as much calcium as was in the oats. A biochemist might suggest that the bird took the calcium from its skeleton. A chicken deprived of calcium in its food will lay soft-shelled eggs after several days, but if it is fed potassium, it will lay normal calcium-shelled eggs. Oats are rich in potassium. And how can a newly-hatched chick's skeleton contain four times more calcium than was in the egg, with the egg shell's calcium content remaining the same? These observations, plus many other experiments by Kervran and others on such diverse organisms as microbes, plants, animals, and humans, have shown that living organisms can indeed change one element into another.[71]

They apparently do this by changes of the atomic nuclei, rather than the outer electrons. The nuclei of elements other than hydrogen contain both neutrons and protons. The number of protons is called the atomic number of that element, which is written as a small number to the lower left of the element's symbol. For example, potassium has an atomic number of 19 (written ${}_{19}K$), hydrogen is one (${}_{1}H$), and calcium is twenty (${}_{20}Ca$). If you take a potassium atom and add a hydrogen atom, you get a calcium atom (${}_{19}K + {}_{1}H = {}_{20}Ca$). Calcium can also be formed by adding oxygen to magnesium (${}_{12}Mg + {}_{8}O = {}_{20}Ca$), or by adding carbon to silicon (${}_{14}Si + {}_{6}C = {}_{20}Ca$). Or you can remove hydrogen from calcium to get potassium, or combine sodium and oxygen (${}_{11}Na + {}_{8}O = {}_{19}K$).[72] This may be partly how soil microbes and some plants can enrich poor soil with certain nutrients and why rotations and companion crops work so well. Kervran believes microorganisms are of utmost importance in

[71] *Ibid.*, p. 286–288.
[72] L. C.Kervran, *Biological Transmutations*, 1972.

correcting soil imbalances, while overuse of synthetic N-P-K fertilizers leads to imbalances and unhealthy crops.[73]

Fertilizing systems. We are all familiar with the conventional system of fertilizing crops based on what nutrients agronomists say the crops remove from the soil, with emphasis on N-P-K, nitrogen, phosphorus, and potassium. It is sometimes called the balance-sheet system . . . putting into the soil what the plants take out. Most farmers use it. Essentially they are feeding their plants, not the soil. Usually with synthetic, purified, immediately available "chemical" fertilizers. It's like a shot in the arm—or better yet, like a massive transfusion! Generally too much fertilizer is put on all at once and it can upset the soil's delicate balance of nutrients and living organisms. It is one of the causes of the many soil problems today (see Chapters 1, 3, and 4). It attempts to control and manipulate nature far too much.

There are other philosophies and systems of fertilizing that are much more in tune with ecology and nature's laws. They vary somewhat in their accepted practices, but there is broad overlap, since they are all working with the same unyielding laws of nature.

1. *Organic agriculture*. First there is the school of thought based on the pioneering work of Sir Albert Howard and perpetuated by J. I. Rodale. Howard was a British horticulturist who in 1924 founded the Institute of Plant Industry at Indore, India. He developed a method of composting agricultural and city wastes called the Indore method. He found that compost helped grow robust, disease-free crops. He saw that the increasing problems of impoverished soils and human disease were related and could be eliminated by proper care of the soil, especially recycling of organic wastes and fostering beneficial soil organisms. His books, *The Waste Products of Agriculture: Their Utilization as Humus* (1931), *An Agricultural Testament* (1940), and *The Soil and Health* (1947) expounded his views.

Right from the start (1931), the "agricultural establishment" gave Howard's ideas a hostile reception. The importance of humus was discredited, and the growing "scientific" technology of N-P-K and frontal assault on pests were said to be superior. But Howard persevered and spread his "gospel" all over the world. In the United States, J. I. Rodale of Emmaus, Pennsylvania, caught the spark. He turned a run-down farm into a showplace using Albert Howard's principles. In 1942, he began publishing *Organic Farming and Gardening* magazine. Many people were receptive to his message. Rodale began a research organization

[73]P. Tompkins & C. Bird, *ibid.*, p. 297.

and published numerous books. The original magazine has split into *Organic Gardening* and *The New Farm*, plus others that are health-oriented, carried on by his son, Robert Rodale.

The basic philosophy of organic agriculture is that fertile soil is maintained by replenishing it with compost instead of chemical fertilizers. The use of commercial fertilizers and highly toxic pesticides should be avoided. There are different degrees of "organic" farmers. The purists absolutely refuse to use any man-made fertilizers or pesticides, while many others may use small amounts of some less harmful fertilizers and/or small doses of insecticides or herbicides if needed. But for the most part, organic farmers rely on recycled organic matter and natural rock fertilizers for fertility, on natural resistance and natural enemies for pest control, and on cultivation for weed control.

One problem many farmers have is trying to "make the switch" from chemical-intensive farming to organic methods. If they quit chemicals "cold turkey," a disaster often results. The soil takes several years to recover from abuse. The Rodale research organization, now called the Regenerative Agriculture Association, is doing considerable research to refine switchover methods.

Also, much Rodale research, plus the experience of thousands of organic farmers, clearly shows that organic farming can be profitable, as well as raising good yields (average to above-average) of high quality crops and healthy animals. Even a number of state universities and the USDA are starting to sit up and take notice, with a leading USDA soil physicist being assigned to the Rodale Research Center in 1984.

2. *Eco-agriculture.* Another "voice crying in the wilderness" is Charles Walters, Jr., editor and publisher of *Acres U. S. A.* Coming from a rural Kansas background, Walters became active in the politically and economically oriented farmers movements during the 1960's, publishing books and articles along the way. In 1972, he began *Acres U. S. A.,* which has blossomed into a widely varied forum for both standard and "off-beat" non-toxic agricultural methods, farmers' alcohol production, and parity-based economics. The farming emphasis is on ecology and healthy plants and animals, so in general the ideas presented in Walters' publications dovetail with those of Rodale and associates, but many other subjects are covered, such as the use of humates, seaweed, and weak energies. Much of the material published in the earlier years was the legacy of the late Dr. William A. Albrecht, who for many years expounded ecologically sound agricultural methods.

3. *Reams and the RBTI.* If anyone could cause bewilderment and head-scratching among agronomists and soil scientists, it was Dr. Cary Reams. He developed a system of soil fertilization that contradicts many

of the establishment's beliefs. But it is based on research of the 1920's and 30's, and it generally gives excellent results. Some of his claimed yields include 20 tons of alfalfa per acre, 100 bushels of soybeans per acre, 90 bushels per acre of wheat, 20 tons per acre of cabbages, 20,000 quarts of strawberries per acre, and so on.[74]

Reams was originally an engineer and in the course of solving problems for clients, became interested in health and agriculture. Through reading the works of early researchers and advising farmers and horticulturists in the Southeast and elsewhere, he developed what he called the Reams Theory of Biological Ionization (RBTI). This theory contains some ideas that conflict with chemistry and agricultural textbooks and is difficult for someone with a background in "standard science" to understand because Reams defined many terms differently than the scientists do. Dr. Reams and his "disciple," veterinarian Dr. Dan Skow, taught their methods in seminars and in booklets and a newsletter written by Skow. Reams also applied his theory to animal and human health and taught his dietary principles in seminars. He died in August 1985.

Some of the principles of the RBTI are as follows. The theory is based on the idea that all occurrences involve some type of magnetic or electrical charge. The surface of the earth has magnetic lines of energy running from the south magnetic pole to the north. This type of magnetic and electrical energy (through ionized particles) is what makes plants grow. The electrical potential of the soil can be easily measured by a conductivity meter; Dr. Reams called this measurement the soil's "energy" or "ergs" (equal to the standard units of micromhos).

Dr. Reams called some nutrients (ions) cations (with a positive charge) and others anions (with a negative charge). He defined an anion as the smallest measurable unit of energy in existence (measured in units called millhouse units; an anion can have from 1 to 499 millhouse units of energy). A cation is defined as the second smallest unit of energy, ranging from 500 to 999 millhouse units of energy. Reactions ("resistance") between anions and cations in the soil produce energy for plant growth. Anionic elements are calcium [contrary to standard chemistry], potassium, chlorine, sodium, nitrate nitrogen, and sometimes oxygen and hydrogen. All other elements are cationic, including ammonium nitrogen.

A preponderance of anionic soil nutrients causes plants to produce vegetative growth (stems and leaves), while cationic elements cause flowering and seed production. Thus, by regulating the balance of soil nutrients, it is possible to control the time of flower-set of a "seed crop," or to keep a "leaf crop" from going to seed (such as hay or lettuce).[75]

[74]D. L. Skow, *The Farmer Wants to Know*, 1979.
[75]*Ibid.*, *Acres U. S. A.*, September 1984, p. 22–26.

There is more to the Reams theory, but that will suffice. Most of the practices Dr. Reams recommended are compatible with organic methods (such as recycling organic matter, fostering soil organisms, and use of some rock fertilizers), but there are some differences. Reams recommended using some of the synthetic "chemical" fertilizers, but only ones that are compatible with soil life and in smaller amounts. He didn't hesitate to recommend using certain insecticides in low dosages, but did not like herbicides. He made use of foliar feeding to help a crop along and emphasized crop quality as measured by sugar content.

4. *Bio-dynamic agriculture.* A somewhat different system of soil stewardship is called bio-dynamics. Much of what is done is similar to organic agriculture, but bio-dynamics has a deeper philosophy behind it.

Bio-dynamic agriculture began when Rudolf Steiner, a German philosopher, gave a series of lectures to a group of farmers in 1924. Steiner had developed a philosophical view of life called anthroposophy ("wisdom of man"), based on the ideas of the well-known German philosopher Johann von Goethe. Anthroposophy is a "path of knowledge that strives to lead the spiritual in man to the spiritual in the universe," according to Steiner. Living organisms are believed to have a spiritual nature. The bio-dynamic system of agriculture encompasses the farm as a whole, aiming toward a "farm organism" in harmony with its habitat, its workers, and the market it supplies.[76]

Dr. Ehrenfried E. Pfeiffer studied Steiner's methods and then transplanted them to the United States in 1939. After rejuvenating the "Threefold Farm" in Spring Valley, New York, he started the Bio-Dynamic Association and began publishing the *Bio-Dynamics Quarterly*. Pfeiffer also developed a bacterial compost starter inoculant.

Bio-dynamic principles are rather rigid in what they require the farmer to do. The emphasis is on restoring the soil's humus through supplying specially-made compost. The compost must be made in an exact way from animal (cattle) manure and other organic matter, soil, and small amounts of special plant "preparations" (made by burying certain wild herbs along with specified animal organs for a summer or a year). The fermented preparations are inserted into a compost pile, where they are said to influence the composting process to proceed in the proper direction. Some preparations are also used as either foliar or soil sprays.[77]

Also, the farmer must use a crop rotation which alternates soil-exhausting with soil-building crops. Companion planting is encouraged,

[76] H. H. Koepf, B. D. Pettersson, & W. Schaumann, *ibid.*, p. 14–28.

[77] *Ibid.*, p. 30, 206–224; E. Pfeiffer, *Bio-Dynamic Farming and Gardening*, 1943, p. 44–47.

and a best ratio of pasture and cropland, and a varied environment surrounding the farm are the ideal goal (a harmonious distribution of fields, pasture, forest, lakes, and rivers).

Tests comparing bio-dynamic methods with conventional agriculture have shown crop yields to be comparable and quality and disease resistance to be excellent. In an animal-feeding test, chickens were either fed grain from bio-dynamically fertilized soil or from conventionally fertilized soil. Thirty eggs from each group of chickens were stored. After two months, 4 (14%) of the conventional eggs had spoiled, compared to only 1 (4%) of the bio-dynamic eggs. After six months, 18 (60%) conventional and 8 (27%) bio-dynamic eggs had spoiled.[78]

Although it is fairly popular in Europe, bio-dynamics does not have many followers in the United States, probably partly because it requires considerable hand labor and dedication.

5. *French intensive method.* This is a method of intensive production from a small area of land that has recently been introduced into California from Europe. It requires much hand labor, so is not practical for large commercial operations, but can produce excellent yields of vegetables for family-sized farms. Crops are grown on raised beds, about 154 square feet in size, which have been deeply tilled (24 inches) and highly fertilized. The beds are surrounded by fresh horse manure to protect from spring frosts. Yields run from three to five times greater than with conventional methods.[79]

Animal health. Modern agriculture, like modern medicine, regards animal disease as primarily being caused by an attack of some sort of pathogen, or "germ." And therefore the remedy is to get rid of the critter doing the damage and the disease will disappear. So zap 'em with some poison that will kill the germs before it kills the animal.

As we have said before (Chapter 4), disease germs attack after an animal is under stress and is already subclinically sick. A healthy animal has natural immunity to pathogens in its environment.

Drugs and most medical remedies have side effects. They may knock out germs, but they also can knock the animal for a loop, adding more stress. Doesn't it make more sense to work *with* nature and build up the animal's immune system so it can resist pathogens in the first place, or fight them off if they do get a foothold?

Besides reducing environmental stresses from cramped quarters, poor ventilation, toxic water, wrong temperatures, loud noises, etc., one of the best ways to prevent health problems (and reproductive dif-

[78]E. Pfeiffer, *ibid.*, p. 174–176.
[79]R. Oelhaf, *Organic Agriculture*, 1978, p. 118.

ficulties) is to feed your animals an excellent diet: high quality food grown on top-notch soil. Such food should be full of all the vitamins, minerals, and other health-promoting factors your animals need for total health. And high quality food should not contain toxic substances, such as nitrates or pesticide residues. Spoiled, moldy grain is potentially contaminated with extremely toxic mold-produced poisons (mycotoxins), such as aflatoxin.

For young mammals especially, dosages of the beneficial *Lactobacillus acidophilus* and other species of bacteria can help overcome diarrhea (scours), poor feed utilization, and other health problems. These good bacteria crowd out the "bad" bacteria that may have gotten established in the digestive tract. The *Lactobacillus* can also detoxify harmful chemicals in the intestine.[80]

To make better quality and more nutritious silage, inoculants of several beneficial bacteria and fungi are commercially available. They aid the fermentation process and produce lactic acid, which by lowering the pH, preserves proteins and other nutrients and prevents growth of harmful microbes. The silage is more palatable to the animals and is partly pre-digested by the microorganisms. Such inoculants are also useful for baled hay.

If animals do come down with an infection, there are products available that use natural antibodies—not antibiotics—to fight the germs. A pregnant female produces antibodies in its milk that protect the newborn from any pathogens to which she has been exposed. Production of antibodies can be induced by man by injecting killed bacteria into the mammary glands. Products have been developed in this way especially for dairy cattle, which are under considerable stress from high production.

Non-food uses for crops. You don't have to grow crops for food (or fiber, such as cotton and flax). There are other things that can be done with plant matter, plus the by-products of food plants or animal wastes can also be used. Some of the uses are as old as the hills, while others are recent developments.

Crude oil is a treasure-store of useful chemicals. Not only do we make vehicle fuel and heating and lubricating oils from it, but a whole array of industrial chemicals, including ether, benzene, xylene (solvents), kerosene, asphalt, and paraffin. Natural gas (methane) usually also is found along with oil. Petroleum is essentially "bottled sunshine," made from the remains of ancient vegetation (or possibly animals). It is possible to produce the same chemicals and many more from modern

[80] *The New Farm*, March/April 1984, p. 28–29.

vegetation, to capture the energy of the sun for cheap fuel and chemical by-products. This is sometimes called *biomass conversion*. Plants produce an amazing variety of chemical by-products themselves, and many others can be produced by chemical transformation by man.[81]

First, let's consider the non-food uses of the typical U. S. agricultural crops and organic wastes. Methane gas can be produced by on-the-farm units from anaerobic bacterial digestion (a 20-day process) of manures or other farm or city wastes. The methane can be converted into electricity or used as fuel. The remainder of the manure can still be recycled into the soil. The California Energy Commission estimated that if all animal manures produced in that state were converted into energy, the savings would equal 27 million barrels of oil a year, or $750 million dollars.[82]

Ethyl alcohol (ethanol) can be produced by fermentation by yeast of any starch, sugar, or cellulose-containing plant matter. Easily available materials include cereal grains (corn, wheat, barley, rye, rice, sorghum, etc.), sugarcane, sugar beets, Jerusalem artichokes, potatoes, as well as old newspaper, garbage, and processing wastes from canneries, dairies (whey), and the paper industry. Spoiled flour or moldy grain work fine. Butyl alcohol (butanol) can be cheaply produced by bacterial fermentation of starchy materials.[83] The alcohol can be used as a vehicle fuel or fuel additive (to produce "gasohol"), although ethanol must be water-free, requiring an additional process. Butyl alcohol can be made into rubber. Or the alcohol can be used for other industrial applications. Farmer-produced alcohol is nothing new. It was common in the 1930's and 1940's (and was a valuable part of the World War II support). But it has been suppressed by the big oil companies and is still fighting an uphill battle.[84] The left-overs from alcohol processing (stillage) can be used as a nutritious animal feed.

Diesel fuel or fuel additive can be made from sunflower seed oil or soybean oil, as well as other vegetable oils.[85] There are also many species of plants not ordinarily grown as crops which could be developed for biomass conversion. Scientists are researching them for possible commercial production of fuels, oils, waxes, resins, rubber, and industrial chemicals. The by-products of processing could be used for animal feed and paper manufacture. Included are such common "weedy" plants as sumac, milkweed, goldenrod, quackgrass, ragweed, thistle, ironweed,

[81] *Science*, 1981, vol. 212, p. 1465–1471; vol. 213, p. 513–517.
[82] *Acres U. S. A.*, May 1982, p. 14.
[83] *Acres U. S. A.*, January 1982, p. 17–19, 26.
[84] *Acres U. S. A.*, July 1983, p. 30–31.
[85] *Acres U. S. A.*, December 1981, p. 18–19; June 1983, p. 15.

resin weed, and spurge, as well as a number of desert and tropical plants (see also the section on new crops earlier in this chapter).[86] The beauty of cultivating such plants is that they will grow in poor and dry soils, and thus will not interfere with ordinary crop production.

Land use. Efficient and proper land use is only a matter of common sense. The earth's surface differs from place to place in topography, soil type, climate, and native vegetation. It only makes sense to use land for purposes that are in harmony with the local topography, climate, and natural resources.

But in all too many instances, that has not been the case. Parts of the western prairies have recently been purchased at low pasture-land prices, plowed, and one or two years' profitable wheat crops grown before the fragile topsoil blows away. You would think the dust bowl days of the 1930's would have been an object-lesson, but not where greed is involved. Growing irrigated field crops in the low-rainfall western United States, with depletion of underground aquifers or surface rivers is another example of short-sighted foolishness. Cities and freeways sprawl out over prime farmland. Erosion-prone row crops are grown on hillsides, with no cover crop being grown over the winter. Protective windbreaks are uprooted to give fields large enough for giant tractors and irrigation rigs. Woods and fence rows are gobbled up, eliminating bug-eating wildlife.

Ideal agricultural land use should suit the local conditions and re-

[86]*Acres U. S. A.*, June 1983, p. 15–16.

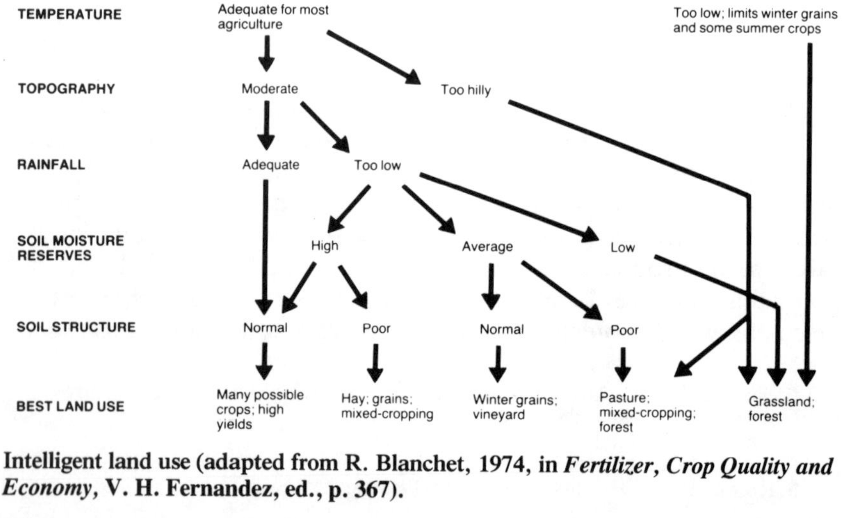

Intelligent land use (adapted from R. Blanchet, 1974, in *Fertilizer, Crop Quality and Economy*, V. H. Fernandez, ed., p. 367).

sources. Uplands and sloping land should be used for pasture and hay crops, with row crops on flatter lowlands. Rows should be tilled around the contour, not up and down a slope. Crops using high soil moisture should not be grown in low-rainfall areas. Land that will only support a forest should be left in forest; wild trees can be profitably harvested for wood or paper, or a nut or fruit orchard can be planted. Housing developments belong on hillsides or wooded land, not on prime farmland. The accompanying chart gives common-sense land uses for temperate climates.

Efficient systems. Next, let's put together all the pieces and make a complete picture. A picture of a much more effective, and more ecologically sound system than our present food production, processing, marketing, and distribution system. Dr. Barry Commoner calls our technologically-oriented agribusiness system "an ecological disaster . . . a main contributor to the ecological crisis."[87] There are a number of people and organizations with excellent ideas for better ways of doing things.

Writing in the book *Sustainable Food Systems*[88], Gil Friend observed that the old biological methods of agriculture were abandoned, not because they didn't work, but because they were out-competed economically by technology based on cheap energy and pesticides. With higher foreign oil prices, energy-intensive technology no longer has an advantage. He believes biological agriculture needs to be productive and sustainable, and conserve resources. There should be limited tillage (but only with no pesticide use), legume-based rotations, and integrated pest management (use of non-toxic methods first). He takes a "whole systems" approach, with a diversified agro-ecosystem (many different crop and animal species, each complementing the others). Productive capacity should be optimized by long-range planning. Soil fertility should be built up by organic matter and microbial action; if outside fertilizers are needed, relatively insoluble ones should be used so that they break down slowly in the soil.

An energetic man named Wes Jackson has set up a practical environmental research center called The Land Institute in the rolling hills of central Kansas, where he and his associates are developing a system of sustainable agriculture. Jackson is against tillage and monoculture in agriculture as being too destructive to fertility and causing erosion. He recommends a "bio-technical fix" to solve agriculture's problems. As described in his book, *New Roots for Agriculture* (1980), this includes a

[87] *The Closing Circle*, 1971, p. 146.
[88] D. Knorr, ed., 1983, p. 28–38.

system of polyculture (many crop species) of perennial species (that live more than one year), thus making tillage unnecessary. Promising perennial plants include fruit and nut trees for food and fuel; perennial species of soybeans and corn (the ones we now grow are annuals), millet, and sunflowers for grain; and native prairie grasses for forage. Corn could also be processed into alcohol to supplement petroleum-based fuels, with the left-overs being used for animal feed.

Jackson envisions future utopian "solar villages" of about 250 population scattered across the country, surrounded by about 128 small farms (about 160 acres). These are average figures for central Kansas; for more fertile land, farms would be smaller, and for less productive land, they would be larger. There would also be larger regional cities of seldom more than 40,000 population. The farms would be self-sustaining as much as possible. Energy would come from solar heating, wind to generate electricity and pump water, and crop alcohol for fuel. Houses would be well-insulated, partly underground. Uplands would be used for pasture grasses, legumes, and sunflowers, with livestock grazed in rotation on different fields (spreading their manure as they go). Lowlands would be devoted to relatively small acreages of monoculture crops (corn, wheat, barley, rye, oats), orchards, woodlot (for supplementary heating), and gardens. Only a small amount of machinery would be needed: a 45 horsepower tractor, a hay baler, and a small multiple-harvesting combine. Seldom-used machinery (planter, fertilizer spreader, chisel plow) could be rented in the village. Land would not be owned by individuals, but could be passed on to future generations. Land use would be under the control of a board of non-farming elders in the village (at least one-half of which had once been farmers).

Another visionary is Gwynn Garnett, Virginia-based grower of high quality biologically-grown beef. Using a background of post-World War II Director of Food and Agriculture in Germany and food market development in several countries, Garnett has developed "systems of small biofarms." The small farms, of 10 to 80 acres each, are coordinated into "modules" of 400 to 800 farms each. Each module has two to four "biofarm exchanges," modified farm supply stores, which provide farmers with needed supplies, technological inputs, and information, and which supervise food production, quality, marketing, employment, services, and community activities. Such a module in Virginia is already paying farmers 10 to 30 percent over the market price for high quality food, and their production costs are one-fourth to one-third less than with conventional methods. Food is marketed through regular supermarket chains from Washington, D. C. to Boston under a special label emphasizing high nutrients and no toxic residues. Other outlets include restaurants, processors, and co-ops.[89]

Rodale Press, Inc., has formed a research and planning organization called The Cornucopia Project, devoted to regenerating the U. S. food system first, then the global food system. A study of the U. S. food system revealed that it is at a critical point, with its foundation crumbling. Topsoil is being lost at the rate of 4.8 billion tons per year by erosion, plus about 3 million acres per year are being lost to urban and highway development. Increasing use of pesticides and chemical fertilizers has contaminated soil and water and led to health problems and resistant pests. Rising production costs have caused serious economic troubles for farmers. Transportation (mainly trucking) costs to move food to consumers were over $16 billion in 1980. Most northeastern U. S. states "import" about 3/4 of their food, from an average distance of 1300 miles. The entire food system could collapse from several interrelated causes.

The Cornucopia Project recommends that farmers reduce their debt

[89]*Acres U. S. A.*, April 1984, p. 30–32; G. Garnett, personal letters and literature, Garnett Coordinated Biofarm Systems, Remington, Virginia, May 1983.

to manageable levels (by diversified farming, reduced chemical usage, developing on-the-farm energy and fertilizer sources, and direct marketing). Consumers should grow some of their own food and buy locally-grown food. The food industry should give preference to farmers using sustainable methods, do less processing and packaging, and encourage consumption of locally-produced food. Local governments can encourage and facilitate local production, marketing, and processing. The federal government should end subsidies to non-sustainable agriculture and strengthen programs for soil conservation. A long-range food plan for the country is urgently needed, with emphasis on regional self-reliance and sustainable farming methods.[90]

There are still other people and organizations, in this country and others, who have their own solutions to our food system problems. It is interesting how similar the various ideas are. They stress self-sustaining, ecologically sound agriculture with small-sized farming units. Pretty much just the opposite of the present system!

What will happen?? Now that we've seen what other people think, what do I think will be the future of agriculture—for this country and for the world?

I foresee a bright future for agriculture. Abundant harvests and people leading happy, productive lives.

But not before some bad news. Things will get worse before they get better. The programs and policies of this present system—and the powers behind them—are so deeply entrenched and have so much momentum, that nothing a relatively weak handful of reformers can do will stop them.

The bankers and politicians will continue to keep us under their thumb. The scientists will cram their technology down our throats. Farms will continue to get bigger and "more efficient." Small family farms will continue to go bankrupt. The complaints and pleas of desperate farmers will be ignored.

Someday, and not too many years hence, the bubble will burst—and the entire economic and technological system will collapse. It's inevitable. The signs of impending collapse are all around us. Increasing bank failures, large companies going bankrupt or merging. Ballooning balance-of-payment deficits and national debt. Rotting inner cities with sickening crime, immorality, and drugs. It's all going to tumble down around our ears, with some pretty disastrous short-term consequences.

But then, when people *finally* wake up and realize that a system

[90]*Empty Breadbasket?*, 1981.

based on greed and taking more from the earth than you give back just can't work, things will get better—much better.

I can see a future where nearly everyone will have a small piece of land to call their own—an agrarian society. They will grow most of their own food—vegetables, fruit, and grain—and will have a few cattle, some chickens, and perhaps a few goats or sheep. Food will be grown without synthetic chemicals and will be tasty and nutritious. Wastes will be composted and recycled to the soil. Wind and solar energy will provide much of the needed energy. Technology will be non-polluting.

The relatively small number of merchants and manufacturers will live in well-planned, well-run small towns. Most people will learn the skills necessary to make many of their tools, furniture, and clothing. Families will be nearly self-sufficient. Life will be relaxed, prosperous, and joyous. Family members will be close; crime and immorality will vanish.

There will be no such thing as "third-world" countries and famines, for eventually all nations will learn to live a better way—a *much* better way.

There's a revolution coming in agriculture. Which revolution do you want?

GLOSSARY

adsorb—accumulation of molecules on the *surface* of a material
aerobic—with oxygen (or air) present (see anaerobic)
anaerobic—without oxygen (or air) (see aerobic)
anion—a negatively-charged ion (see cation)
annual—a plant that lives only one year (see perennial)
biomass—the total amount of living organisms of one or more species in a certain area
cation—a positively-charged ion (see anion)
chelate—a water soluble organic compound that can hold a metal cation
chemical—any particular kind of matter
colloidal—having very small particle size (less than 0.002 mm, or 1/5000 inch), but larger than most molecules; with a large surface area
compound—a material composed of two or more elements
ecosystem—a total ecological system composed of all the living organisms plus the non-living environment
foliar—referring to leaves
inorganic—chemical definition: chemical compound not containing carbon; ordinary definition: not produced by living organisms (see organic)
ion—an electrically charged atom or group of atoms
metabolic—referring to the chemical activities of living cells or organisms
microbe—same as microorganism
microorganism—a microscopic organism
molecule—the smallest particle of a chemical compound, composed of two or more atoms or elements
monoculture—growing one species or variety of crop in an area (see polyculture)
organic—chemical definition: chemical compound containing carbon; ordinary definition: referring to or produced by living organisms (see inorganic)

organism—any living thing, whether microbe, plant, or animal
pathogen—a disease-causing microorganism
perennial—a plant that lives more than one year (see annual)
pH—a measure of acidity or alkalinity, with pH 7 being neutral, pH 0 to 7 acid, and pH 7 to 14 alkaline
polyculture—growing a mixture of crop species in an area (see monoculture)
rhizosphere—the area immediately next to a plant's roots
species—a certain kind of organism
symbiotic—two (or more) species of organisms living together in close association and having an effect on one another
synergistic—two (or more) agents (chemicals, actions, etc.) having a greater effect together than the sum of their separate effects

APPENDIX A

Common Fertilizer Materials

Because of the varying nutrient needs of crops, different levels of existing soil nutrients, different characteristics of fertilizer materials, and the possibility of harming crops and soil by wrong fertilizer use, it is important to know something about the fertilizer materials you use. If a fertilizer label does not specify what materials are contained, it is wise to find out from the supplier or manufacturer, because some materials are definitely less desirable than others (see Chapters 3 and 5).

In the following table, important characteristics of commonly used fertilizer materials are given (compiled from various sources).

Material	% Nitrogen	% Phosphorus	% P_2O_5	% Potassium	% K_2O	Other Characteristics
alfalfa hay	2-3	0.09-0.26	0.2-0.6	1.7-2.65	2.0-3.2	slowly available
ammonia						
• anhydrous (gas)	82-82.2					used as liquid (injected); kills soil organisms & roots nearby; hardens soil; later turns soil acid
• aqua (liquid)	16-25					
ammonium nitrate	32.5-34					supplies two types of nitrogen; nitrate can leach; turns soil acid; often used in liquid fertilizer mixtures
ammonium polyphosphate						
• dry	10	15	34			
• liquid	15	25-27.3	57-62			
ammonium sulfate	20-21					23.7% sulfur; turns soil acid
ammonium thiosulfate	12-18.9					used as liquid; 26–43.3% sulfur
basic slag		3.5-12	8-27.5			29–32% calcium; also supplies several trace elements; makes soil alkaline
bean straw or residue	1.2	0.1	0.25	1.04	1.25	slowly available
blood meal, dried blood	8-15	0.7-0.9	1.5-2.0	0.5-0.8	0.6-1.0	slowly available
bone meal	0.2-4.5	5.3-15.0	12-34			slowly available; 20–25% calcium; also supplies small amounts of magnesium and sulfur
brewer's waste	0.9-3.2	0.04-0.2	0.1-0.4	0.4-1.7	0.5-2.0	slowly available
calcite (see calcium carbonate)						
calcium carbonate						32–40% calcium; makes soil alkaline
calcium hydroxide						46–54% calcium; rapidly available; makes soil alkaline
calcium magnesium carbonate						22% calcium, 8–20% magnesium; slowly available; makes soil alkaline

Material	% Nitrogen	% Phosphorus	% P_2O_5	% Potassium	% K_2O	Other Characteristics
calcium nitrate	15-17					19–21% calcium; readily available; nitrate can leach; makes soil alkaline
calcium oxide						60–71% calcium; very strong, can kill crops and soil organisms; makes soil alkaline
calcium sulfate						22–23% calcium, 15–20.5% sulfur
castor bean pomace	6	0.5-0.7	1.2-1.5	0.4	0.5	slowly available
colloidal phosphate (see rock phosphate)						
cottonseed meal	5.7-7.0	0.9-1.3	2-3	1.24	1.5	slowly available
diammonium phosphate (DAP)	16-21	20-23.7	46-54			can injure seedlings (high ammonia); later turns soil acid
dolomite (see calcium magnesium carbonate)						
epsom salts (see magnesium sulfate)						
feather meal	12					slowly available
fish meal, fish emulsion	4-11	1.3-3.5	3-8	0-2.5	0-3	slowly available; also supplies small amounts of calcium, magnesium, and sulfur
granite dust				3-5	3.6-6.0	slowly available; also supplies magnesium and iron
grass hay, grass clippings	0.1-5.0	0.01-0.9	0.025-2.0	0.1-2.5	0.12-3.0	slowly available
greensand (glauconite)		0.44	1	5-7	6-8.4	slowly available; also supplies magnesium and iron
gypsum (see calcium sulfate)						
hoof and horn meal	12.5	0.77	1.75			slowly available
hops, spent	0.6-5.7	0.09-1.5	0.2-3.4	0.008-2.16	0.01-2.6	slowly available

Material	% Nitrogen	% Phosphorus	% P_2O_5	% Potassium	% K_2O	Other Characteristics
kalium potash (see potassium chloride)						
kelp (see seaweed)						
K-mag (see sulfate of potash-magnesia)						
langbeinite (see sulfate of potash-magnesia)						
leather dust, leather meal	0.4-12	0.04-0.18	0.1-0.4	0.017-0.25	0.02-0.3	slowly available
leaves	0.5-1.0	0.02-0.035	0.05-0.08	0.25-0.5	0.3-0.6	slowly available
lime, dolomitic (see calcium magnesium carbonate)						
lime, high calcium (see calcium carbonate)						
lime, hydrated (see calcium hydroxide)						
lime, magnesium (see calcium magnesium carbonate)						
lime, quick (see calcium oxide)						
lime, slaked (see calcium hydroxide)						
lime, unslaked (see calcium oxide)						
magnesium sulfate						9.6–10.5% magnesium, 11–14% sulfur
malt fiber	0.6-5.7	0.09-1.5	0.2-3.4	0.9-1.8	1.1-2.2	slowly available
manure						
• cattle	0.5-0.7	0.09-0.2	0.2-0.45	0.4-0.66	0.5-0.8	
• horse	0.4-0.7	0.09-0.2	0.2-0.5	0.37-0.66	0.45-0.8	slowly available; should be composted or incorporated shallowly for best results
• poultry	1.1-1.7	0.9-0.6	1.0-1.3	0.4-0.8	0.5-1.0	
• sheep	1.0-2.0	0.15-0.44	0.35-1.0	0.4-1.6	0.5-2.0	
• swine	0.5	0.13	0.3	0.37-0.54	0.45-0.65	
marl (see calcium carbonate)						similar to high calciumn lime, but lower calcium content because of clay impurities

Material	% Nitrogen	% Phosphorus	% P_2O_5	% Potassium	% K_2O	Other Characteristics
mono-ammonium phosphate (MAP)	11-13	8.7-21	20-48			also supplies small amounts of calcium, magnesium, and sulfur; later turns soil acid
monocalcium phosphate (see superphosphate and triple superphosphate)						
muriate of potash (see potassium chloride)						
peanut hulls	1.5-3.6	0.05-0.2	0.12-0.5	0.66	0.8	slowly available
peanut meal	7.2	0.66	1.5	1.0	1.2	slowly available
peat	2.7					slowly available
phosphoric acid		23-24	52-54.5			used as liquid; turns soil acid
potassium chloride				49.8-51.5	60-62	47.6% chlorine; high salt index; can kill soil organisms and roots
potassium magnesium sulfate (see sulfate of potash-magnesia)						
potassium nitrate	13.0-13.8			36.5-38.7	43.8-46.6	nitrate can leach
potassium sulfate				39.8-44	48-53	17.6–18% sulfur
poultry litter	0.18-0.28	0.03-0.1	0.07-0.23	0.04-0.15	0.05-0.18	slowly available
rock phosphate						
• hard		11-17	25-39	0.3	0.36	30–36% calcium; also supplies several trace elements; slowly available
• soft		8-10	18-23	0.26	0.32	19% calcium; also supplies several trace elements
sawdust	0.05-0.9	0.004-0.2	0.01-0.5	0.03-1.16	0.04-1.4	slowly available
seaweed	0.6-0.9	0-0.2	0-0.5	3.3-10.8	4-13	slowly available; also supplies several trace elements
sewage sludge						
• digested	1.5-2.0	0.6	1.4	3.3-15.0	0.4-18.0	slowly available; also supplies small amounts of calcium, magnesium, and sulfur
• activated	2-6	1-3	2.2-7.0	0-0.8	0-1.0	
soybean meal	7	0.5-0.7	1.2-1.6	1.2-1.9	1.5-2.3	slowly available; also supplies small amounts of calcium, magnesium, and sulfur

Material	% Nitrogen	% Phosphorus	% P_2O_5	% Potassium	% K_2O	Other Characteristics
straw (small grain)	0.6	0.09	0.2	0.9	1.1	slowly available
sulfate of potash-magnesia				18.3-19.2	22-23	22.7–23% sulfur, 11.2% magnesium
sulfur (elemental, flowers of)						88–100% sulfur; slowly available; turns soil acid
sul-po-mag (see sulfate of potash magnesia)						
superphosphate, ordinary		7-9.6	16-22			18–21% calcium, 11–14% sulfur
tankage						
• animal	6-9	2.6-4.4	6-10	0-1.2	0-1.5	11.5% calcium; also supplies small amounts of magnesium and sulfur; slowly available
• garbage	2-3	0.44-1.3	1-3	0.8-1.6	1-2	slowly available; also supplies small amounts of calcium, magnesium, and sulfur
tobacco stems	1.5	0.2	0.5	4.15-8.3	5-10	slowly available; also supplies small amounts of calcium, magnesium, and sulfur
triple superphosphate		19.8-23	52.7			14.3% calcium
urea						
• dry	42-46					can damage seedlings (high ammonia); do not leave on surface; often used in liquid fertilizer mixtures
• liquid	20					
winery pomace	1.5	0.66	1.5	0.66	0.8	slowly available
wool wastes	7.5					slowly available

Trace Elements

A trace element (micronutrient) is an element that plants need for normal growth and development, but only in very small, trace amounts. Some trace elements are abundant in soils, but in certain regions, one or more may be deficient. Wrong fertilizer balance, pH, or soil conditions may also cause trace element deficiencies. See Chapter 5 of this book and Chapters 4, 11, and 12 of *The Rest of the Story . . . about Agriculture Today* for more details.

Many trace elements are found in the above-listed materials, especially the rock fertilizers and organic materials. The purified salt fertilizers and synthetic materials have few if any. Following is a list of some materials used to supply trace elements. Those that are sulfates and chelates are readily available to plants, while most of the others will give long-term availability.

Remember that plants only need small amounts of trace elements, and some of them are toxic in too great amounts. Trace elements should not be applied unless a reliable soil test indicates they are needed.

Element	Material	% of Element
boron	borax	11
	boric acid	17
	sodium pentaborate	18
	sodium tetraborate	13–20
	boron frits	10–17
	Solubor	20.5
copper	copper sulfate	25–35
	basic copper sulfate	13–53
	copper ammonium phosphate	32
	cupric oxide	75
	cuprous oxide	89
	copper chloride	16.9
	copper acetate	32
	copper oxalate	40
	copper chelates	7.5–13
	copper frits	3–50
iron	ferrous sulfate	19–20
	ferric sulfate	20–23
	ferrous ammonium sulfate	14
	ferrous carbonate	42
	iron ammonium phosphate	29
	iron ammonium polyphosphate	22
	iron chelates	5–15
	iron frits	10–40
manganese	manganese sulfate	23–28
	manganese carbonate	31
	manganese oxide	63
	manganous oxide	41–68
	manganese chloride	17
	manganese chelates	5–12
	manganese frits	10–35
molybdenum	sodium molybdate	39
	ammonium molybdate	54
	molybdenum trioxide	66
	molybdenum frits	1–30
zinc	zinc sulfate	36
	zinc carbonate	52–56
	zinc oxide	50–80
	zinc chloride	50
	zinc chelates	6–14
	zinc frits	10–30

APPENDIX B
Composting

Since antiquity farmers have been recycling organic wastes, either by applying fresh manure, straw, etc., to the soil or else by accumulating a pile of assorted refuse (a compost heap) which gradually rots to form compost. According to *The Complete Book of Composting* (J. I. Rodale, editor, 1960, p. 19), compost is decomposed organic matter, and composting is the process of making humus.

We have covered the value and functions of humus, especially in Chapters 2 and 5. If soil is short of humus, there will likely be problems with fertility, soil structure, erosion, weeds, crop diseases, and pests, not to mention crop quality.

One of the best ways to build up soil humus without the danger of toxicity from anaerobic decomposition of fresh organic matter, is to turn the organic matter into humus first, then apply it to the soil—namely composting.

Besides the benefits of the humus produced, composting also benefits the farmer by (1) destroying weed seeds, disease pathogens (both plant and animal diseases), and parasites, (2) reducing fly breeding, and (3) reducing the volume of manure or other organic wastes that has to be handled (to 1/4 or less of the original volume), plus (4) reducing obnoxious odors.

What happens? During the process of composting, fresh organic matter, which ideally should contain a ratio of about 30 parts carbon to 1 part nitrogen, is attacked by a "volunteer army," a variety of bacteria, fungi, and actinomycetes. These microbes use the organic matter for food and rapidly multiply into millions. They gradually decompose the originally complex molecules in the organic matter (proteins, fats, cellulose, hemicellulose, starches, and many others) into simpler materials, some of which they use to build their own cells. But after they die, the changed organic substances are able to be used by plants for their growth, in the form of simple organic molecules and inorganic ions (sulfate, nitrate, ammonium, calcium, potassium, etc.). Much of the original carbon content of the organic matter is consumed by the microbes for energy and "exhaled" as carbon dioxide, leaving the final compost with a carbon-to-nitrogen ratio of about 10 or 20:1.

As the microbes begin their work, the temperature inside the compost pile rapidly rises, up to 150–160°F. This is the best temperature for rapid composting in the early stages, and this is why weed seeds and disease pathogens are killed.

After a few weeks the temperature of the compost pile gradually drops, as other species of microbes take over decomposition. For about the first two months, the microbes need air (oxygen) to grow, so proper aeration of the compost pile is important. After that, anaerobic species finish the composting. The compost is "ripened" or "finished" (after three or more months) when the temperature of the pile reaches nearly the outside temperature, no ammonia odor can be detected, and the material has a dark color and rich "earthy" smell. The compost should have no large fibers or stems left of the original organic matter.

Practical notes. For farm composting on a large scale, the following notes may be helpful:

1. *Materials.* Almost any organic material can be composted, including animal manures, green manure (fresh plants), crop residues, straw, old hay, seaweed and kelp, weeds, peat moss, tree bark, city leaves and grass clippings, sawdust and wood chips, brewery and cannery wastes, molasses residue, soybean and cottonseed meals, linseed and peanut meals, hulls and shells (of oats, rice, buckwheat, cocoa beans, and cottonseed), tobacco stems, castor pomace, papermill and sewage sludges, shredded newspapers, garbage, slaughterhouse wastes (bloodmeal, bone meal, hoof and horn meal, and tankage), fish meal, and feathers.

Ideally, your raw materials should include both plant and animal materials, since plant matter is high in carbon (carbohydrates) and animal matter contains more nitrogen. At least 10–20% of the compost pile should be animal manure, which supplies bacteria, enzymes and growth promoters, and nitrogen. Low nitrogen material such as sawdust or wood chips should make up less than 1/3 of the pile. If not enough nitrogen is available, a suitable amount of nitrogen fertilizer can be added (about 15 lbs. actual nitrogen per ton), such as fish meal, bone meal, blood meal, cottonseed meal, soybean meal, urea, or ammonium sulfate (the latter causes acid conditions; add an equal amount of lime). If you are composting mainly animal manures, slaughterhouse wastes or fish meal (high nitrogen), a considerable amount of plant matter should be added to supply carbon, absorb moisture, and provide aeration. Plant matter should not be too coarse or the pile will dry out too easily. It is desirable to mix in at least 10–20% soil, which traps ammonia (prevents nitrogen loss), provides beneficial microbes, and buffers acid pH. Sometimes unwanted soil can be obtained from construction sites (sand and heavy clays are not good). Some old compost can be used in place of the soil. If a soil test shows you need any trace elements, they are more effectively added to a compost pile rather than to the soil, since in the composting process they are converted into forms more readily available to plants (chelates).

The pH of the composting material should be slightly acid or around neutral. A small amount of lime (2 lbs. per cubic yard) will stimulate bacterial action and neutralize acids (soil is also a good acid neutralizer). The same amount of soft rock phoshate is a good addition and will enrich the phosphorus content of the compost. Some powdered kelp (seaweed) or humates will help the crops after the compost is applied, if not the composting process itself. The pH should not become very alkaline, however, because this inhibits microbial action and allows nitrogen to escape in the form of ammonia. If you smell a strong ammonia odor, the pH is too high (some ammonia is normal at first, however). Sprinkling a little superphosphate (0-20-0) or humates on fresh barn manure daily will help conserve nitrogen (ammonia) that rapidly escapes otherwise.

2. *Location of pile.* Composting is best done in shallow (two feet) pits or on concrete slabs, but any nearly level piece of ground will do. It is best to have well-packed clay to reduce seepage of liquids into the soil. It is most convenient to either locate the compost piles near the source of materials (manure) and a water supply (if you need to dampen the piles), or else near the fields you intend to use the compost on. If you live in a rainy climate, try to orient the windrows so they do not dam up rainwater. Also, cover the pile with fibrous materials such as straw or leaves to help shed rainwater.

3. *Size of pile.* The most convenient form of pile in farm-scale composting is a long windrow. In order to heat up properly and retain moisture, the pile should be 4 to 5 feet high and 9 to 12 feet wide at the bottom, gradually tapering to a peak or mound at the top. Do not make it larger or air will not be able to get inside.

4. *Building the pile.* Depending on your farm operation, you may build a windrow quickly if a lot of material is available all at once, or slowly, such as by daily cleaning of barn manure. If the windrow is built gradually, build it to full height at one end and add to it. A small front-end loader or tractor loader can be used, or some farmers throw manure out of a manure spreader.

Before starting a windrow, it is good to lay down a 1 to 1 1/2 foot layer of absorbent litter, such as straw, corn stalks or cobs, sawdust or wood chips, etc., to absorb nutrient-rich liquids that seep out of the pile, and to aid in ventilation. When making the pile, be sure to mix the various ingredients as thoroughly as possible.

5. *Starting decomposition.* Some compost experts recommend adding a special bacterial inoculant to insure an abundant supply of the best species of microbes. If the wrong combination of species is present, composting will be slow and wasteful of nutrients, so if you are in a hurry or the pile is not composting properly, an inoculant is a good investment.

In many cases, the raw materials (manure, soil, etc.) will already have a good assortment of microbes, and no inoculant is necessary. Once you have made some compost, saving a little from an old pile to inoculate the new pile is an excellent method. Home-made inoculants can also be made by soaking some compost, leaf mold or humus, or topsoil in several gallons of water for an hour or more and spraying or pouring this "compost tea" on the new pile (the better contact it has with the material inside the pile, the faster composting will begin).

6. *Moisture.* The material should be between 40 and 60% moisture for best composting, although slower action will occur from 25 to 75% moisture. If the material is soaking wet or juicy, it is too wet; mix in some absorbent plant matter (straw, ground corn cobs, shredded newspaper, etc.). If it is dry, add water. If the weather is dry and the pile may dry out, covering it with a 6-inch thick layer of straw will retain moisture.

7. *Aeration.* For the first eight weeks or so, the pile needs to be well aerated to compost properly. If the temperature has reached the danger level (see next section), turning the pile will also aerate it. If the temperature is lower, poking holes into the center of the pile with a pipe or post will aid aeration if it is needed. However, if enough loose plant matter was added and if the pile is not too moist, this is unnecessary. And if there is too much ventilation the pile may not heat up or may dry out.

A good method of ventilation for large operations is to build the windrows over perforated pipes or hoses connected to an air blower. This allows larger windrows to be built.

8. *Temperature.* When first built, the pile should rapidly heat up to well over 100°F. If it gets above 160°F, it should be turned (see next section) to prevent wrong decomposition. The temperature can be measured by sticking a thermometer (perhaps fastened onto a stick) into a hole poked into the pile. If the pile is "steaming," the temperature is probably about right or possibly near the danger level. If a pile refuses to heat up, it may be too wet or too dry, or poorly aerated. In the winter, covering the pile with a thick layer of straw, etc., will help insulate it.

The temperature should stay at the high levels for about two or three weeks. The high temperatures are valuable in killing weed seeds, pathogens, and parasites. After a few weeks, the temperature will gradually cool down as the compost is "ripening."

9. *Turning*. During the first approximately eight weeks, the pile must be watched carefully and turned whenever the temperature gets above 160°F for best results and fastest composting. Sometimes this may mean turning every few days. If you cannot spend so much time, composting will still occur, but may be greatly delayed and with waste of nutrients. The pile should be turned at least twice, about three and five weeks after being made.

There are fairly expensive compost turning machines on the market, and for large composting operations, they are necessary. If you have a talent for building machinery, you can make your own. For smaller operations, a compost windrow can be turned with a front-end loader or else loaded onto a manure spreader and re-piled beside the old windrow.

The idea in turning is to mix up the material and if possible put what used to be on the inside on the outside, and vice versa. This insures that everything gets composted at the same speed and that weed seeds and pathogens get killed. Don't worry about saving the bottom layer of litter (see section 4). If the material begins to dry out after turning, add water (spray the outside of the pile with a hose).

10. *The finished product*. As composting proceeds, the volume of the material will shrink considerably. Depending on original materials, from 20 to 60% of the original volume will disappear as it is "eaten" by microbes. Up to 50% of the weight can be lost.

The final result, compost, should be dark brown, crumbly, moist, rich-smelling humus, without much recognizable structure (no large stalks, stems, or other remains of the original organic matter). It should be full of beneficial microorganisms, waiting to go to work when added to your soil. It should have an abundance of plant nutrients, from major elements to trace elements to hormones and growth promoters. The percent of nitrogen will be greater than in the original organic matter (remember, the carbon-to-nitrogen ratio goes from about 30:1 to 10:1), and if nitrogen-fixing microbes were active, it can be a goldmine of free nitrogen. I have found composted manure to have two or three times more readily available nitrogen than the original manure.

If finished compost must be stored for a long time, it is best to cover it to prevent leaching of nutrients or drying out. Cover with wet burlap, tarps, or plastic sheets. Finished compost should be turned and watered occasionally to prevent undesirable anaerobic decomposition.

11. *Application*. To obtain maximum benefit from compost, it should not be let dry out, but spread and quickly incorporated into the

soil (within hours). Spreading on a cloudy or rainy day is good.

Compost can be added to soil at any time without harm, except that applying just before planting a crop that doesn't require much nitrogen (for example, small grains) may cause problems. A convenient time to apply is after removal of one crop and before tillage for the next crop.

The rate of application can vary considerably. Good results have been obtained with anywhere from 1/2 to 10 tons per acre. It partly depends on the nutrient needs of the next crop, and if you are trying to increase soil humus content, improve tilth (soil structure), or fight dry climate or drought, then use higher application rates. After soil is improved, much lower rates can be used.

An imaginative way of applying compost has been used by some southwestern U. S. farmers who use ditch irrigation. A pit is dug in a main feeder irrigation ditch, filled with compost, and flooded. The flow of water carries plant nutrients and beneficial microorganisms to plants. The pit should be about four feet deep (not over six feet) and can hold about 150 tons of compost. Supplementary side dressing with compost can also be done. Yields and crop quality on irrigated desert land can be tremendously improved (*The Complete Book of Composting*, 1960, p.462–466).

Index

A

B

C

D

E

F

G

H

I

L

M

N

O

P

Q

R

S

T

U

V

X

Y

Z